TUTORIAL CHEMISTRY TEXTS

10

Thermodynamics and Statistical Mechanics

JOHN M. SEDDON & JULIAN D. GALE

Imperial College of Science, Technology and Medicine, University of London

ISBN 0-85404-932-1

A catalogue record for this book is available from the British Library

Published by The Royal Society of Chemistry, Thomas Graham House, Science Park,
Milton Road, Cambridge CB4 0WF, UK
Registered Charity No. 207890
For further information see our web site at www.rsc.org

Typeset in Great Britain by Wyvern 21, Bristol
Printed and bound by Polestar Wheatons Ltd, Exeter

Preface

In this text we aim to cover the basic results of thermodynamics and statistical mechanics which are needed by undergraduate chemistry students. We cover in some detail the properties of ideal gases, since this model system allows analytical solutions to be developed, and the results derived can be used more widely. However, we do not here cover topics such as the properties of non-ideal gases, liquids and solids, nor the thermodynamics of solutions and electrochemical cells.

Thermodynamics is a subject which many students find confusing and difficult to grasp. In part this is due to the large number of thermodynamic quantities and relations they have to master. In this text we have attempted to introduce each of these in a logical order, then develop them in a clear and coherent way to try to guide the student, so they will gain confidence in applying thermodynamic analysis to chemical problems. A further difficulty some students encounter is with the level of mathematics required, and we have tried to keep this as simple as possible.

We have followed closely the IUPAC recommendations for nomenclature [I. Mills *et al.*, *Quantities, Units and Symbols in Physical Chemistry*, 2nd edn., Blackwell Scientific, Oxford, 1993], even though some usage may be unfamiliar to many students. Thus, for example, we write $\Delta_{vap}H_m$ for the molar enthalpy of vaporization; note that the subscript m is frequently omitted, the units indicating whether or not a molar quantity is intended.

This book is based closely upon current first-year thermodynamics and second-year statistical mechanics courses given to undergraduate chemistry students at Imperial College, London. We are deeply indebted to our course predecessors, Dr. Garry Rumbles, Professor Michael Spiro and Professor Dominic Tildesley, for their sterling efforts in developing these courses and course notes, and for their invaluable and extensive advice. Any errors or points of confusion in the text are of course entirely our responsibility. We are grateful to Martyn Berry for his careful reading of the manuscript.

John Seddon and Julian Gale
London

TUTORIAL CHEMISTRY TEXTS

This series of books consists of short, single-topic or modular texts, concentrating on the fundamental areas of chemistry taught in undergraduate science courses. Each book provides a concise account of the basic principles underlying a given subject, embodying an independent-learning philosophy and including worked examples. The one topic, one book approach ensures that the series is adaptable to chemistry courses across a variety of institutions.

TITLES IN THE SERIES

Stereochemistry *D G Morris*
Reactions and Characterization of Solids *S E Dann*
Main Group Chemistry *W Henderson*
d- and f-Block Chemistry *C J Jones*
Structure and Bonding *J Barrett*
Functional Group Chemistry *J R Hanson*
Organotransition Metal Chemistry *A F Hill*
Heterocyclic Chemistry *M Sainsbury*
Atomic Structure and Periodicity *J Barrett*
Thermodynamics and Statistical Mechanics *J M Seddon and J D Gale*
Mechanisms in Organic Reactions *R A Jackson*
Quantum Mechanics for Chemists *D O Hayward*

FORTHCOMING TITLES

Molecular Interactions
Reaction Kinetics
Basic Spectroscopy
X-ray Crystallography
Lanthanide and Actinide Elements
Maths for Chemists
Bioinorganic Chemistry
Chemistry of Solid Surfaces
Biology for Chemists
Multi-element NMR

Further information about this series is available at www.chemsoc.org/tct

Orders and enquiries should be sent to:
Sales and Customer Care, Royal Society of Chemistry, Thomas Graham House, Science Park, Milton Road, Cambridge CB4 0WF, UK

Tel: +44 1223 432360; Fax: +44 1223 426017; Email: sales@rsc.org

Contents

Fundamental Constants

Planck constant	h	6.626×10^{-34} J s
Boltzmann constant	k_B	1.381×10^{-23} J K^{-1}
Avogadro constant	N_A	6.022×10^{23} mol^{-1}
Gas constant	$R (= N_A k_B)$	8.315 J K^{-1} mol^{-1}
Faraday constant	F	9.649×10^{4} C mol^{-1}
Speed of light in vacuum	c	2.998×10^{8} m s^{-1}
Elementary charge	e	1.602×10^{-19} C
Rest mass of electron	m_e	9.109×10^{-31} kg
Rest mass of proton	m_p	1.673×10^{-27} kg
Atomic mass constant	m_u	1.661×10^{-27} kg
Acceleration of gravity	g	9.807 m s^{-2}
Standard atmosphere	atm	101 325 Pa (N m^{-2})
Standard pressure	$p^{\ominus}$	10^{5} Pa (N m^{-2})

1 Introduction

Thermodynamics and quantum mechanics are the two fundamental pillars of chemistry. The latter is mainly concerned with the microscopic properties of atoms or molecules, whereas classical thermodynamics concerns the macroscopic or bulk equilibrium properties of matter. The link to molecular properties is the subject of statistical mechanics, which is dealt with in the second half of this text. Thermodynamics underpins all chemical process and reactions. On the one hand, it is based on quite abstract and subtle concepts; on the other hand, it is an extremely practical subject, dealing with questions such as: "What is the equilibrium constant for this reaction and how does it vary with temperature?"

In this introductory text, we will see that thermodynamics – like quantum mechanics – depends on a small number of basic ideas and postulates, and that everything else follows in a straightforward way if the underlying principles are grasped. Thermodynamics is never wrong; if a process appears to break the Laws, it means that our analysis is wrong or incomplete. We will look at simple examples to illustrate the basic principles involved. It should be noted that the mathematics required is generally rather simple, once you are clear about what you are trying to calculate and how you are going to go about it.

Thermodynamics is essentially concerned with the conservation of energy, and with energy transfer, either in an organized form (work), or in a chaotic, disorganized form (heat). It predicts the spontaneous direction of chemical processes or reactions, and the equilibrium states of chemical systems. However, it does not deal with the *rates* of processes or reactions: this is the subject of chemical kinetics. Thermodynamics arose in part out of the need in the 19th century to improve the design of steam engines. As the subject evolved it was found that the results were perfectly general, applying equally to all chemical and physical systems. Many of the underlying concepts are easiest to understand when applied to purely mechanical systems, such as the expansion of a gas in a cylinder. Where appropriate, we will use such systems as illustrative examples and in problems.

In this chapter we will introduce the four Laws of Thermodynamics, then explore their consequences for chemical processes. In order to do this, we will first have to make some formal definitions.

Aims

By the end of this chapter, you should be able to:

- State the Laws of Thermodynamics
- Define what is meant by open, closed, isolated, adiabatic and diathermic systems
- Understand exothermic and endothermic processes

1.1 The Laws of Thermodynamics

There are four Laws of Thermodynamics. Somewhat confusingly, the first of these is known as the Zeroth Law. It is concerned with the definition of temperature and thermal equilibrium, and may be stated as:

"There is a unique scale of temperature."

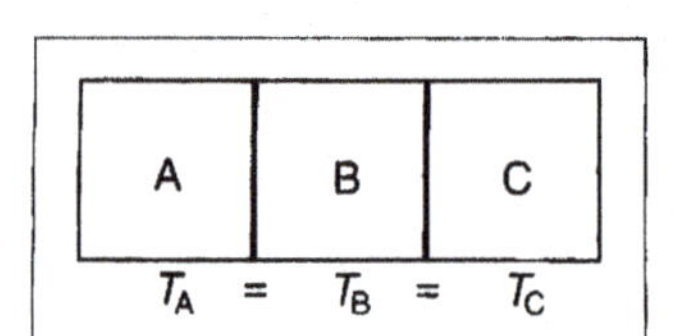

Figure 1.1 Thermal equilibrium

Thus if body A is in thermal equilibrium with body B, and B with C, then A is also in thermal equilibrium with C (Figure 1.1).

The First Law is concerned with the conservation of energy, and may be stated as:

"The energy of an isolated system is constant."

This means that energy can be neither created nor destroyed, only transferred between systems, or between a system and its surroundings.

The Second Law is concerned with the spontaneous direction of processes, determined by changes in the entropy S (introduced in Chapter 4). One way of expressing it is as follows (see Figure 1.2):

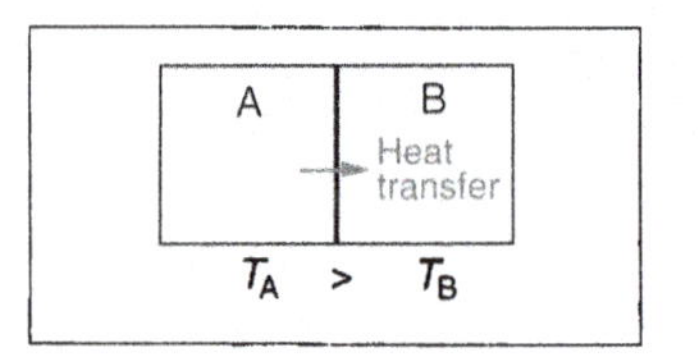

Figure 1.2 Direction of spontaneous heat flow

"When two systems are brought into thermal contact, heat flows spontaneously from the one at higher temperature to the one at lower temperature, not the other way round."

There are many equivalent statements of the Second Law, such as:

- Heat cannot be completely converted into work for any cyclic process, but work can spontaneously be completely converted into heat.
- Spontaneous changes are always accompanied by a conversion of energy into a more disordered form.

- The entropy S of an isolated system increases during any spontaneous change or process.

The Third Law is also concerned with entropy, and may be stated as follows:

> *"All perfect materials have the same entropy S at $T = 0$, and this value may be taken to be $S = 0$; at higher temperatures, S is always positive. It is impossible to cool any system to $T = 0$."*

1.2 Definitions

Before proceeding with exploring the implications and applications of the Laws, it is necessary to make a number of definitions (see Figure 1.3):

System: region of chemical interest, *e.g.* a reaction vessel, characterized by properties such as T, p, V, composition, *etc.*

Surroundings: region outside the system, sometimes where we make our measurements, and separated from the system by a boundary.

Open system: both energy and matter can be exchanged between the system and its surroundings.

Closed system: only energy can be transferred, either as work, or by heat transfer.

Isolated system: neither energy nor matter can be exchanged.

Adiabatic system: the system is thermally isolated, and so heat transfer cannot occur, although work can be performed on or by the system.

Diathermic system: a system for which heat transfer is possible.

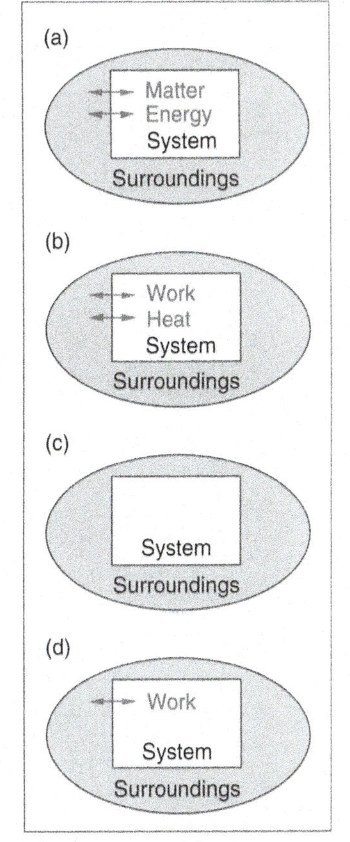

Figure 1.3 Types of system: (a) open; (b) closed; (c) isolated; (d) adiabatic

1.3 Exothermic and Endothermic Processes

An exothermic process is defined as:

> *"A process which* releases *energy as heat."*

An example would be the combustion of hydrogen. If the system is *adiabatic*, the heat released stays in the system, *raising* its temperature. If the system is *diathermic*, and maintained at a constant temperature by a thermal reservoir, the heat released flows from the system into the surroundings (the thermal reservoir).

An endothermic process is defined as:

> *"A process which* absorbs *energy as heat."*

An example would be the vaporization of water. If the system is adiabatic, the heat is absorbed from the system, *lowering* its temperature. If the system is diathermic, heat flows in from the surroundings in order to maintain the system at the same temperature as the surroundings.

Summary of Key Points

1. *Laws of Thermodynamics*
   Temperature scale; conservation of energy; spontaneous direction of heat flow and chemical processes; entropy at absolute zero.

2. *Systems and surroundings*
   Open, closed, isolated, adiabatic and diathermic systems.

3. *Types of thermal process*
   Exothermic and endothermic processes.

Further Reading

E. B. Smith, *Basic Chemical Thermodynamics*, 4th edn., Oxford University Press, Oxford, 1990.

P. W. Atkins, *Physical Chemistry*, 6th edn., Oxford University Press, Oxford, 1998.

R. A. Alberty and R. J. Silbey, *Physical Chemistry*, 2nd edn., Wiley, New York, 1996.

R. G. Mortimer, *Physical Chemistry*, Benjamin Cummings, Redwood City, Calif., 1993.

G. Carrington, *Basic Thermodynamics*, Oxford University Press, Oxford, 1994.

D. A. McQuarrie and J. D. Simon, *Molecular Thermodynamics*, University Science Books, Sausalito, Calif., 1999.

D. Kondepudi and I. Prigogine, *Modern Thermodynamics*, Wiley, Chichester, 1998.

H. DeVoe, *Thermodynamics and Chemistry*, Prentice Hall, Upper Saddle River, NJ, 2001.

K. E. Bett, J. S. Rowlinson and G. Saville, *Thermodynamics for Chemical Engineers*, Athlone Press, London, 1975.

2
The First Law

Having introduced the Laws of Thermodynamics in the previous chapter, we now need to learn how to apply them to chemical systems. We will begin this task by considering energy, and energy transformations.

Aims

In this chapter we will examine different forms of energy contributing to the *internal energy* of a system, and explore the difference between *heat* and *work*. By the end of this chapter you should be able to:

- Calculate the work involved in various processes
- Show that both the work and the heat flow depend on the *path* followed during the process
- Understand the difference between reversible and irreversible processes
- Understand that the change in internal energy does not depend on the path followed, but only on the initial and final states
- Define the difference between a *path function* and a *state function*

2.1 Internal Energy

The internal energy U of a system is the total energy it contains as a result of its physical state, that is, under specified conditions of T, p, V, *etc*. The First Law states that the internal energy of a system is constant unless it is changed by work, w, or by heat transfer, q:

$$\Delta U = w + q \tag{2.1}$$

A large, macroscopic change in a quantity X is denoted ΔX; a small, incremental change in X is denoted dX.

U does not depend on how that state was reached; we say that U is a state function. The value of U depends on the amount of matter in our system; if we double the mass, we double U. We say that U is an extensive property of the system. However, the molar value, $U_m = U/n$, does not depend on the amount of matter present: it is an intensive property of the system. In microscopic terms (*i.e.*, at the molecular level), U is equal to the total sum of the energy levels of the atoms or molecules making up the system, weighted by their probabilities of being occupied.

Worked Problem 2.1

Q An *ideal* gas is one which obeys the ideal gas law $pV = nRT$. What types of motion contribute to its internal energy (neglect any contribution from excited electronic states)?

A The internal energy of an ideal gas of atoms (*e.g.* argon), with N atoms of mass m and speed v_i, consists solely of the translational (kinetic) energy of the atoms:

$$U = U_{trans} = \sum_{i=1}^{N} \frac{1}{2} m v_i^2 = \frac{1}{2} Nm < v^2 > \quad (2.2)$$

The symbol Σ denotes the sum of the terms to the right of the symbol. For example:

$$\sum_{i=1}^{4} x_i = (x_1 + x_2 + x_3 + x_4)$$

The symbol $<x>$ denotes the average value of the quantity x.

where $<v^2>$ is the mean-square speed of the atoms. For an ideal gas of *molecules* (*e.g.* N_2 at low pressure) there will be additional contributions from vibration and rotation:

$$U = U_{trans} + U_{rot} + U_{vib} \quad (2.3)$$

These three contributions to the internal energy will be calculated in detail in Chapters 11 and 12.

2.2 Heat

Heat flow may be defined as:

> *"The transfer of energy due to a difference in temperature between the system and its surroundings."*

- If $T_{system} < T_{surroundings}$, q is *positive*, and heat flows *into* the system, *raising* its internal energy U.
- If $T_{system} > T_{surroundings}$, q is *negative*, and heat flows *out of* the system, *lowering* its internal energy U.

For example, if a sealed flask of nitrogen gas is placed in a hot oven, heat will flow into the flask ($q > 0$), raising the temperature and the internal energy of the gas (by increasing the populations of the higher energy levels of the N_2 molecules).

2.3 Work

Work may be defined as:

> *"A process which could be used directly to move an object a certain distance against an opposing force."*

In general, the work w is defined by the integral of the vector dot product of the force $\boldsymbol{F}$ and the small incremental displacement d$\boldsymbol{s}$:

$$W = \int \boldsymbol{F} \cdot \mathrm{d}\boldsymbol{s} \tag{2.4}$$

The dot product between two vectors $r_1 = (x_1, y_1, z_1)$ and $r_2 = (x_2, y_2, z_2)$ is: $r_1 \cdot r_2 = (x_1x_2 + y_1y_2 + z_1z_2)$

If the force is constant, and in the opposite direction to the direction of motion of the object, this simplifies to:

$$W = -F\Delta s$$

Where Δs is the distance the object has moved.

Worked Problem 2.2

Q Calculate the work required to lift a weight (mass) m through a height Δh.

A Assuming that Δh is small, so that g (the acceleration due to gravity) is constant, the opposing force is $-mg$ (see Figure 2.1).

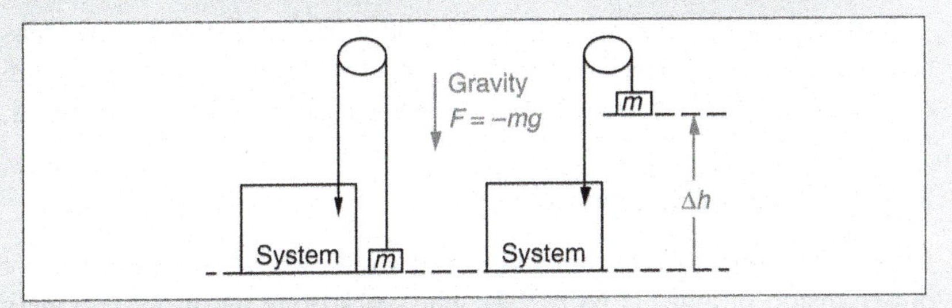

Figure 2.1 Work involved in lifting a weight

The work done *by* the system is then:

$$w = -mg\Delta h \tag{2.5}$$

Note that w is *negative* when Δh is positive; the internal energy U of the system has *decreased.*

If the work is done *on* the system (*e.g.* by letting the weight m fall), then w is *positive*, and the internal energy U of the system is

2.3.1 Types of Work

It is important to understand that there are many different types of work, and these can all change the internal energy of the system (see Table 2.1).

Table 2.1 Examples of different types of work

Type	dw	*Example*
Extension	$f\,dl$	Bond or spring (f = force constant; l = length)
Volume	$-p_{ex}\,dV$	Piston (p_{ex} = external pressure; V = volume)
Surface	$\gamma\,d\sigma$	Liquid surface (γ = surface tension; σ = area)
Electrical	$\phi\,dq$	Battery (ϕ = potential; q = charge)

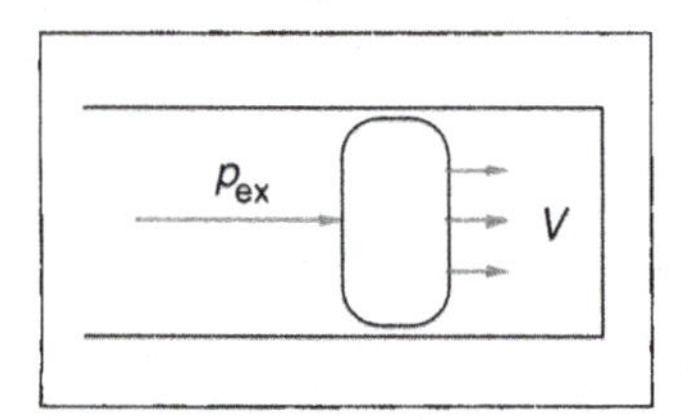

Figure 2.2 Work of compression

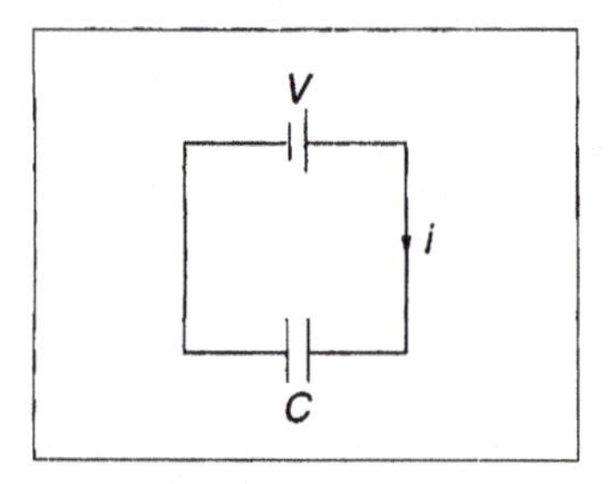

Figure 2.3 Electrical work. The work of charging a capacitor is $q^2/2C$, where q is the charge on the capacitor and C is the capacitance

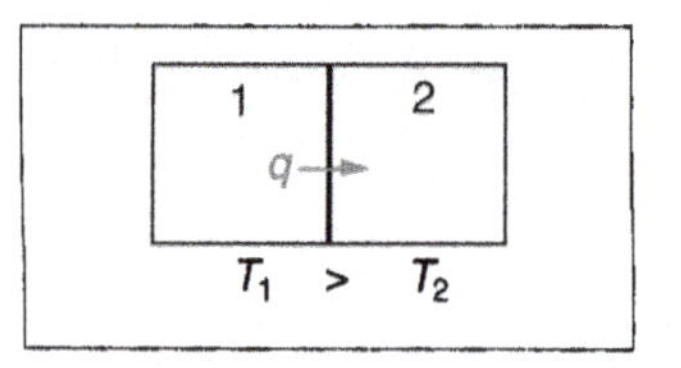

Figure 2.4 Heat flow

It is also important to grasp the difference between work and heat. Work is the transfer of energy due to the *organized* motion of molecules (or atoms, electrons, photons, *etc.*). For example, the atoms in a piston all move together in the same direction to compress a gas in a cylinder (Figure 2.2).

As another example, a battery of voltage V performs electrical work in driving electrons preferentially in one direction along a wire to charge a capacitor (Figure 2.3).

Heat is the transfer of energy between the system and the surroundings due to the *chaotic*, *disorganized* motion (*i.e.* the thermal motion) of molecules (or atoms, electrons, *etc.*). For example, two metal blocks at different temperatures are brought into thermal contact. There is a net heat flow from the higher temperature block to the lower temperature block (Figure 2.4).

Note that metals are good thermal conductors for the same reason that they are good electrical conductors: the availability of conduction electrons which are relatively free to diffuse around inside the metal.

2.3.2 Work of Expansion

As a first example of the calculation of work, we will consider the work of expansion of a gas (Figure 2.5). The work done by the system against the surroundings in moving a piston of area A by a distance dz against an opposing force $F(z) = p_{ex}(z)A$ is:

$$\mathrm{d}w = -F(z)\mathrm{d}z = -p_{ex}(z)A\mathrm{d}z = -p_{ex}(V)\mathrm{d}V \tag{2.6}$$

The system will expand if $p\ (= p_{in}) > p_{ex}$. The negative sign means that for an *expansion* (dV positive), the work is *negative*, *i.e.* the internal energy U of the system *decreases* by carrying out the work on the surroundings. If $p_{ex} > p$, the system will undergo *compression* (dV is negative), and thus the work is *positive*, *i.e.* the internal energy U of the system *increases*. Note that *in both cases* it is the *external* pressure p_{ex} which determines the amount of work done, *not* the system pressure p.

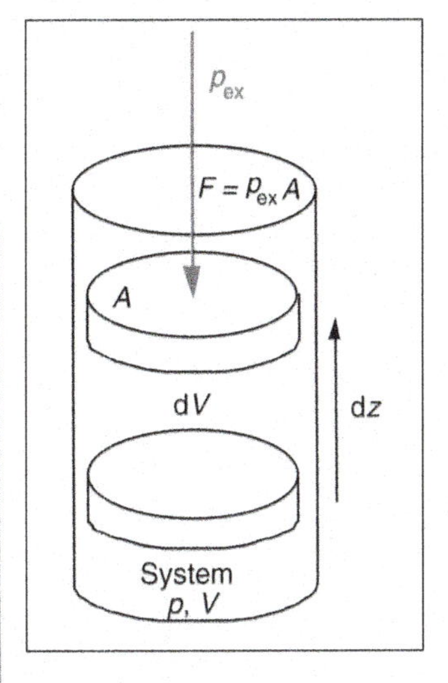

Figure 2.5 Compression or expansion of a gas by a piston

Worked Problem 2.3

Q Calculate the work of free expansion (p_{ex} = 0) of a gas.

A If the surroundings are evacuated and the piston released, the gas will expand. Because there is no opposing force (since p_{ex} = 0), dw = 0 at all stages of the expansion and hence:

$$w_{exp} = \int \mathrm{d}w = -\int p_{ex}\mathrm{d}V = 0 \tag{2.7}$$

Thus no work is done by the system. This is an irreversible expansion.

Worked Problem 2.4

Q Calculate the work of expansion of a gas against a constant external pressure.

A This is a common situation in chemistry. The work is:

$$\begin{aligned} w_{exp} &= \int_{V_A}^{V_B} p_{ex}(V)\mathrm{d}V = -p_{ex}\int_{V_A}^{V_B}\mathrm{d}V = -p_{ex}(V_B - V_A) \\ &= -p_{ex}\Delta V \end{aligned} \tag{2.8}$$

The work is *negative* for ΔV *positive*, *i.e.* work is done *by* the system. The *magnitude* of the work may be illustrated by the use of an indicator diagram, *i.e.* a plot of p_{ex} versus V (Figure 2.6).

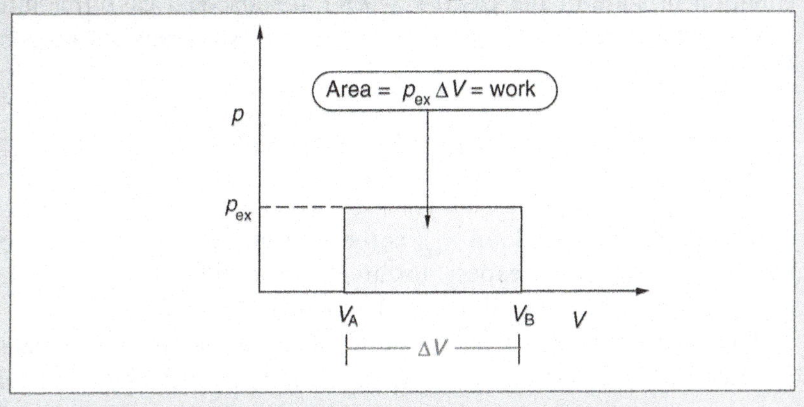

Figure 2.6 Indicator diagram

The magnitude of the work is the area under the curve (but note that the diagram does not give the *sign* of the work). In general, we need to know how p_{ex} varies during the expansion in order to obtain the work, either by evaluating the work integral, or by plotting $p_{ex}(V)$ against V, and measuring the area under the curve.

Isothermal Reversible Expansion

The expansion may be carried out *reversibly* by adjusting p_{ex} to be equal to p (the pressure within the system) at each infinitesimal step in the expansion, thereby keeping the system always in equilibrium (Figure 2.7). The work is then:

A reversible process is one which is carried out in infinitesimally small incremental steps, such that the system remains essentially in equilibrium at all stages during the process.

$$w_{exp} = -\int_{V_A}^{V_B} p_{ex}(V)\mathrm{d}V = -\int_{V_A}^{V_B} p\mathrm{d}V \tag{2.9}$$

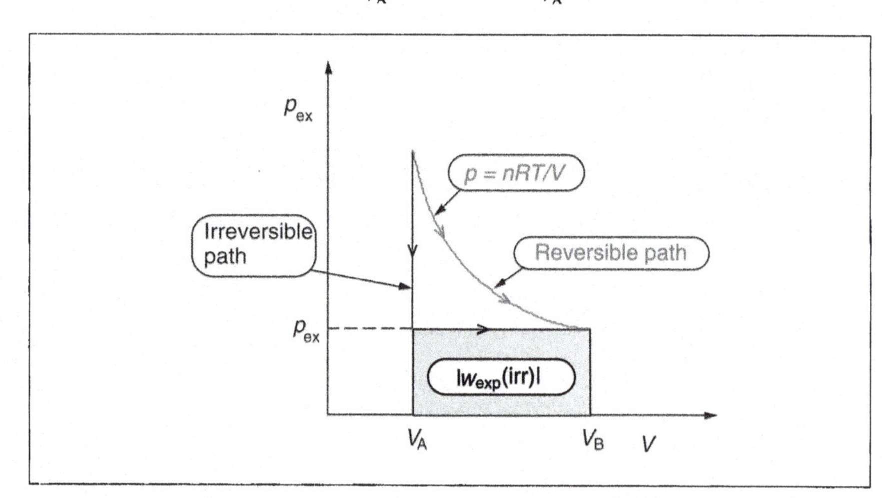

Figure 2.7 Isothermal (T = constant) expansion of an ideal gas.

Note that, to simplify the notation, we leave it implicitly understood that p varies with V. Since the pressure p of the system is given by the equation of state of the gas ($pV = nRT$ for an ideal gas), this integral can be evaluated directly if T is constant, (*i.e.* isothermal). Thus, for an ideal gas:

$$w_{exp} = -nRT\int_{V_A}^{V_B} \frac{\mathrm{d}V}{V} = -nRT\ln\left(\frac{V_B}{V_A}\right) \tag{2.10}$$

The partial derivative $(\partial f/\partial x)_y$ denotes the derivative of function f with respect to the variable x, whilst holding the variable y to be constant.

Note that the magnitude of w_{exp} is the area under the $p = nRT/V$ curve from V_A to V_B. As we expect, for an expansion ($V_B > V_A$), w_{exp} is negative, *i.e.* work has been done by the system.

The internal energy U is constant during an isothermal expansion, since $(\partial U/\partial V)_T$ is zero for an ideal gas (see Worked Problem 2.1). Thus

$\Delta U = (q + w) = 0$, and hence:

$$q = -w_{exp} = +nRT\ln\left(\frac{V_B}{V_A}\right) = +nRT\ln\left(\frac{p_A}{p_B}\right) \quad (2.11)$$

Thus, in the present example, heat q is transferred from the surroundings to the system to balance exactly the work w_{exp} done by the system, and thereby keep the internal energy U of the system constant.

Irreversible Expansion

If the expansion is carried out *irreversibly*, by *suddenly* lowering p_{ex} from p_A to p_B, the expansion occurs against a constant external pressure $p_{ex} = p_B$, and the work of expansion is given by:

$$w_{exp} = -p_{ex}\Delta V = -p_B\Delta V \quad (2.12)$$

Note that this result applies when the system is either adiabatic or isothermal. Thus the work performed in an irreversible expansion is smaller in magnitude than the work of a reversible expansion:

$$|w_{exp}(\text{irr})| < |w_{exp}(\text{rev})| \quad (2.13)$$

This can be seen directly from the areas under the two curves on the indicator diagram (Figure 2.7). This result is an example of a general principle which can be stated as:

The magnitude of a quantity f is denoted $|f|$, the modulus of f. For example: $|-3| \equiv |3| = 3$.

> *"A system operating between specified initial and final states does the maximum work when the process is carried out reversibly."*

This statement is not restricted to work of expansion (pV work), but applies to all kinds of work (chemical, electrical, electrochemical, *etc.*).

Isothermal Compression

We can return the system from V_B to V_A either reversibly or irreversibly, by compression (Figure 2.8).

Reversible Compression

If we gradually increase p_{ex} from p_B to p_A, keeping the system in equilibrium at each step, the system will compress along the same curve ($p = nRT/V$) as for the reversible expansion, and the work (area under

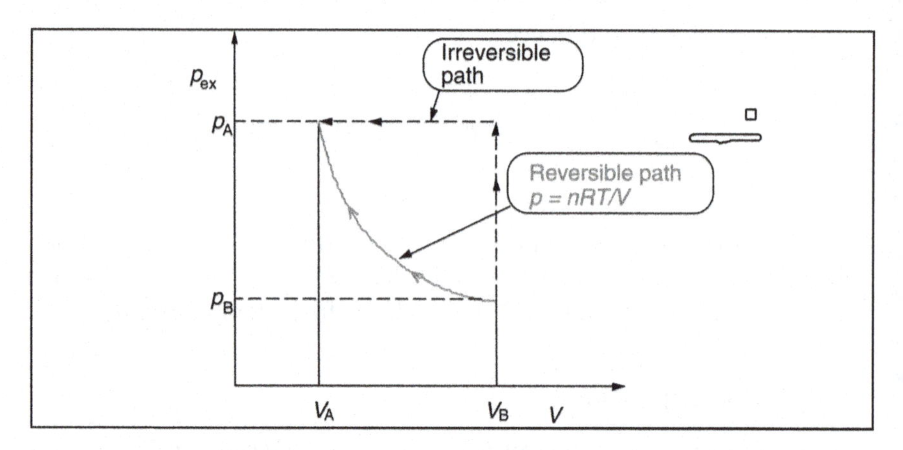

Figure 2.8 Isothermal compression of an ideal gas

the p–V curve) will therefore have the same magnitude (but opposite sign because ΔV is now negative):

$$w_{comp}(\text{rev}) = -w_{exp}(\text{rev}) = +nRT\ln\left(\frac{V_B}{V_A}\right) \tag{2.14}$$

Thus w_{comp} is *positive*, and the same amount of work has been done *on* the system by compression as was done *by* the system during the expansion. Since $\Delta U = 0$ (for an ideal gas), an amount of heat:

$$q = -w = -nRT\ln\left(\frac{V_B}{V_A}\right) \tag{2.15}$$

flows from the system to the surroundings (q is *negative*).

Irreversible Compression

To carry this out, we suddenly increase p_{ex} to p_A (for example, by quickly placing a suitable mass onto the piston, then stopping it at volume V_A). The work on the system is then:

$$w_{comp}(\text{irr}) = -p_{ex}\Delta V = -p_A(V_A - V_B) = +p_A(V_B - V_A) \tag{2.16}$$

and is positive. We see that $w_{comp}(\text{irr}) > w_{comp}(\text{rev})$, and thus the work done *on* the system is *smallest* when the compression is carried out *reversibly*. This is the opposite of the result for the work done *by* the system (*e.g.* during an expansion). It is clear from comparing the indicator diagrams that:

$$|w_{comp}(\text{irr})| > |w_{exp}(\text{irr})|$$

i.e. more work is done *on* the system during an *irreversible* compression than is done *by* the system during an *irreversible* expansion. Since the work done on the surroundings was different for the two different paths between the same states (V_A and V_B), *i.e.* the reversible and the irreversible paths, it is clear that *work is not a state function* (described at the beginning of this chapter). Since the heat transfer is also different for the two paths (to exactly counterbalance the work), the *heat* is also clearly *not* a state function. We refer to work w and heat q as path functions, since their values will depend on the path taken between the initial and final states.

Consider an infinitesimal change in U:

$$\mathrm{d}U = \mathrm{d}q + \mathrm{d}w \tag{2.17}$$

The total internal energy change in taking the system from state A to state B is given by:

$$\Delta U = \int_{U_A}^{U_B} \mathrm{d}U = (U_B - U_A) \tag{2.18}$$

and ΔU is independent of the path taken between A and B (Figure 2.9): the internal energy U is a state function. See Box 2.1.

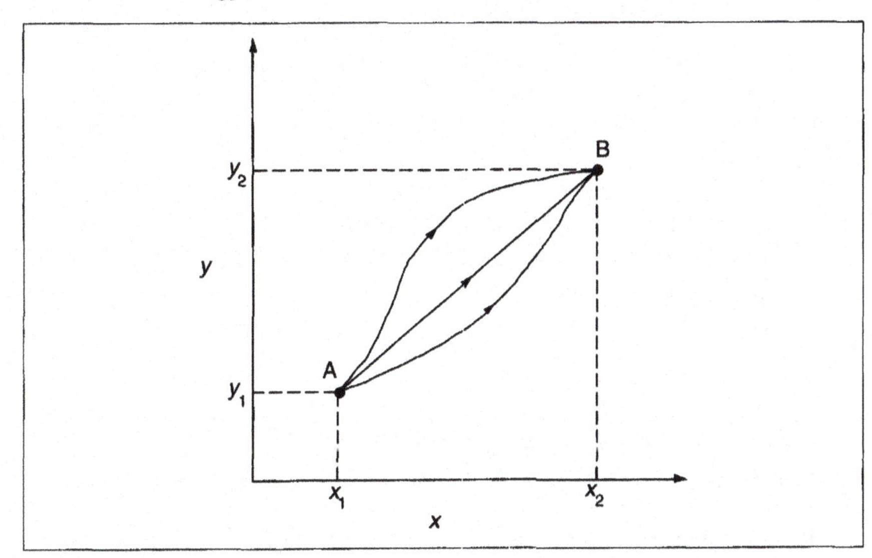

Figure 2.9 For a state function such as $U(x,y)$, ΔU is independent of the path followed from the initial state A to the final state B. The variables x and y are any two of pressure p, volume V and temperature T

Box 2.1 Exact and Inexact Differentials

Mathematically speaking, we can write $\mathrm{d}U$ as an *exact* differential, that is:

$$\mathrm{d}U = \left(\frac{\partial U}{\partial x}\right)_y \mathrm{d}x + \left(\frac{\partial U}{\partial y}\right)_x \mathrm{d}y$$

where $(\partial U/\partial x)_y$ is the partial derivative of U with respect to x at constant y, dx denotes a small incremental change in the variable x, and the variables x and y are any two of pressure p, volume V and temperature T.

Since dU is an *exact* differential, U is a *state* function. On the other hand, dq and dw cannot be expressed in this way: they are *inexact* differentials, which means that q and w are *path* functions. This provides a useful working definition of whether thermodynamic quantities are *state* or *path* functions. We will see later that important quantities such as the enthalpy H, the entropy S and the Gibbs free energy G are all *state* functions. Thus changes in these quantities, ΔH, ΔS and ΔG, depend only on the initial and final states of the system, and are *independent* of the path taken.

Summary of Key Points

1. *The First Law*
   Internal energy; work; heat.

2. *Reversible and irreversible processes*
   Work and heat of expansion and compression.

3. *Path and state functions*
   Work and heat are path functions (dw and dq are inexact differentials), but the internal energy is a state function (dU is an exact differential).

Further Reading

E. B. Smith, *Basic Chemical Thermodynamics*, 4th edn., Oxford University Press, Oxford, 1990, chapter 2.
P. W. Atkins, *Physical Chemistry*, 6th edn., Oxford University Press, Oxford, 1998, chapters 2 and 3.
R. A. Alberty and R. J. Silbey, *Physical Chemistry*, 2nd edn., Wiley, New York, 1996, chapter 2.
R. G. Mortimer, *Physical Chemistry*, Benjamin Cummings, Redwood City, Calif., 1993, chapter 2.
D. A. McQuarrie and J. D. Simon, *Molecular Thermodynamics*, University Science Books, Sausalito, Calif., 1999, chapter 5.

Problems

1. An *adiabatic* system consisting of a thermally insulated reaction vessel with a 10 Ω resistance heater inside is connected to a 20 V power supply, which is switched on for 50 s.
(i) Analyse the change in internal energy ΔU of the system.
(ii) Repeat the analysis for an *isothermal diathermic* system (*i.e.* the system is in contact with a thermal reservoir such as a water bath).

2. The internal energy U of a monatomic ideal gas is $U = \frac{1}{2}Nm\langle v^2 \rangle$, where N is the number of atoms, m is the mass of one atom, and the kinetic theory of gases gives the mean-square speed $\langle v^2 \rangle$ of the atoms as $\langle v^2 \rangle = 3pV/Nm$.
(i) How does U depend on temperature?
(ii) Does U depend on pressure or volume, at constant temperature?

3. A volume of an ideal gas is contained within a cylinder with a frictionless piston at one end. When the internal volume of the cylinder is $V_1 = 1\ \mathrm{dm}^3$, the outward pressure on the piston is $p_1 = 10$ atm. The piston is held stationary by an opposing pressure consisting of 1 atm due to the air outside and 9 atm due to nine weights sitting on the piston (each weight exerts 1 atm pressure). Calculate the work done if:
(i) all of the weights are removed quickly together;
(ii) five of the weights are removed quickly together, the system allowed to equilibrate, and then the remaining four weights are removed quickly;
(iii) the weights are removed one at a time, the system being allowed to come to equilibrium at each step.
Discuss how the *maximum* amount of work could be extracted from the system. Calculate this value.

4. One mole of an ideal gas is carried through the following cycle:

		A		B		C	
	1	→	2	→	3	→	1
V_m/dm^3	22.4		22.4		44.8		22.4
T/K	273		546		546		273

Assuming each process is carried out reversibly:
(i) Calculate the pressure at each state, 1, 2 and 3.
(ii) Name each process, A, B and C.
(iii) Obtain expressions for the heat flow q, the work w, and the internal energy change ΔU for each process.
(iv) Calculate numerical values for q, w and ΔU for the complete cycle.

3

Heat Capacity, Enthalpy and Thermochemistry

Our next task is to learn how to analyse heat flow in chemical systems. This turns out to be the key to deriving the entropy, and hence many of the most important thermodynamic properties.

Aims

By the end of this chapter you should be able to:

- Define the heat capacity C under constant volume, and constant pressure, conditions
- Describe how C varies with temperature, and what occurs at transitions between phases
- Define the enthalpy function H
- Analyse the heats of reactions, and their dependence upon temperature.

3.1 Heat Capacity, *C*

When heat flows into a system, then provided no phase change (such as boiling) occurs, its temperature will increase. The increment dT by which T increases is proportional to the amount of heat flow dq:

$$dT \propto dq$$

The ratio dq/dT is called the heat capacity (units: J K^{-1}):

$$C = \left(\frac{dq}{dT}\right) \tag{3.1}$$

The heat capacity is thus the increment in heat dq required to increase the temperature by an amount dT. C varies for different substances, for different phases (*e.g.* solid, liquid, gas) and in general will vary with temperature and pressure (although it does not for an ideal gas). C is an extensive quantity, but the *molar* heat capacity, $C_m = C/n$, is an *intensive* property. Frequently, the subscript m is omitted; the quoted units then define whether C or C_m is intended. A typical value for C_m is that for liquid water at 25 °C: $C_m \approx 80$ J K^{-1} mol^{-1}.

3.1.1 Isochoric (Constant Volume) Heat Capacity

If the heat is supplied at a constant sample volume (dV = 0), the system can do no pV work, and so (assuming the system does no other kind of work), from the First Law:

$$(\mathrm{d}q)_V = \mathrm{d}U$$

The isochoric heat capacity C_V is then defined as:

$$C_V = \left(\frac{\partial U}{\partial T}\right)_V \tag{3.2}$$

Note, we write the derivative using "∂" symbols to denote that C_V is the partial derivative of U with regard to T, at constant V.

3.1.2 Isobaric (Constant Pressure) Heat Capacity

If the heat is supplied at a constant sample pressure, usually the sample will expand, doing work against the external pressure. Thus not all of the heat will be used to raise the temperature, and a greater heat flow $(\mathrm{d}q)_p$ will be required to raise the system temperature by the same amount as in the constant volume case. Thus the isobaric (constant pressure) heat capacity C_p will normally be greater than C_V. From the First Law (assuming only pV work):

$$\mathrm{d}U = \mathrm{d}q - p_{\mathrm{ex}}\mathrm{d}V$$

Thus:

$$\mathrm{d}q = \mathrm{d}U + p_{\mathrm{ex}}\mathrm{d}V$$

Therefore:

$$C_p = \left(\frac{\partial U + p_{\mathrm{ex}}\mathrm{d}V}{\partial T}\right)_p \tag{3.3}$$

Note that the change in internal energy ΔU is normally less for a given heat flow for an isobaric than for an isochoric process.

3.1.3 Temperature Dependence of C_p

Later, we will see how to calculate enthalpy (ΔH), entropy (ΔS) and Gibbs free energy (ΔG) changes of chemical processes at different temperatures. The key to doing this is the knowledge of how the heat capacity C_p varies with temperature (Figure 3.1).

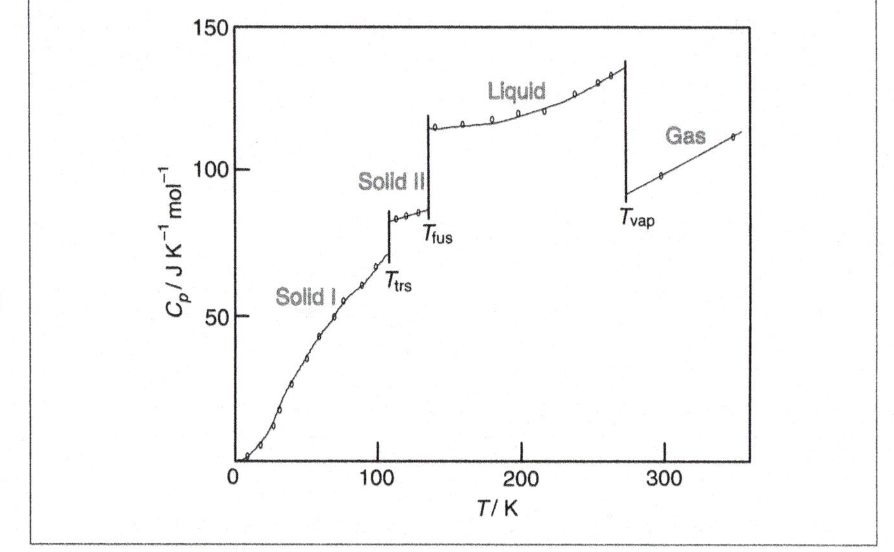

Figure 3.1 Schematic sketch of the variation of C_p with temperature for butane. T_{trs}, T_{fus} and T_{vap} are the solid I–solid II transition temperature, the melting point and the boiling point, respectively (Adapted from E. B. Smith, *Basic Chemical Thermodynamics*, Oxford University Press, Oxford, 1990)

For some systems (*e.g.* monatomic gases such as argon), C_p is independent of temperature. However, for most systems the heat capacity varies in quite a complex way with temperature. Furthermore, at first-order *phase transitions* such as melting or boiling points, the heat capacity tends to infinity ($C_p \rightarrow \infty$) because the heat flow dq does not change the temperature until the latent heat of the transition has been supplied. Within the gas phase, we can use an empirical form for the temperature dependence of C_p:

$$C_p = a + bT + \frac{c}{T^2} \tag{3.4}$$

Values of the constants a, b and c are listed for many elements and compounds in textbooks (*e.g.* see Atkins or Alberty and Silbey in Further Reading).

3.2 Enthalpy

It is useful to define a new thermodynamic quantity, the enthalpy H, such that (when no additional work other than pV work is present):

$$\mathrm{d}H = (\mathrm{d}q)_p \tag{3.5}$$

and:

$$\Delta H = (q)_p$$

This may be achieved by defining the enthalpy as:

$$H = U + pV \tag{3.6}$$

Worked Problem 3.1

Q Show that $\mathrm{d}H = (\mathrm{d}q)_p$.

A Differentiating equation (3.6) for H gives:

$$\begin{aligned}\mathrm{d}H &= \mathrm{d}U + p\mathrm{d}V + V\mathrm{d}p \\ &= (\mathrm{d}q - p_{ex}\mathrm{d}V) + p\mathrm{d}V + V\mathrm{d}p\end{aligned}$$

But $p_{ex} = p$, and we impose constant pressure by setting $\mathrm{d}p = 0$. Hence $\mathrm{d}H = (\mathrm{d}q)_p$, as required.

The expression for the isobaric heat capacity then takes on a simple form in terms of H:

$$C_p = \left(\frac{\partial H}{\partial T}\right)_p \tag{3.7}$$

The importance of the enthalpy function is that the change in enthalpy ΔH is equal to the heat flow q at constant pressure, a condition which often applies in chemistry. Furthermore, since the enthalpy is defined in terms of U, p and V, which are all *state* functions, H is also a state function, and thus ΔH is independent of the path taken between the initial and final states. The subject of thermochemistry relies upon this fact.

It should be noted that for *solids and liquids*, which are nearly incompressible (*i.e.* $\Delta V \approx 0$), $\Delta H \approx \Delta U$. However, for *gases* (*e.g.* an ideal gas, $pV = nRT$), $\Delta H = \Delta U + \Delta nRT$, and if the number of moles of gases in the system changes, $\Delta H \neq \Delta U$.

3.3 Thermochemistry

Consider a reaction (assumed to go to completion):

$$A + B \rightarrow C + D$$

If the initial system has an enthalpy H_1, and the final system has an enthalpy H_2, the enthalpy change upon reaction is:

$$\Delta H = (H_2 - H_1)$$

If ΔH is *negative* ($H_2 < H_1$), heat is *released*, and the reaction is exothermic. If ΔH is *positive* ($H_2 > H_1$), heat is *absorbed*, and the reaction is endothermic. If the reaction is carried out under *adiabatic* conditions, there is no heat flow between the system and the surroundings, and the enthalpy change ΔH causes a change of temperature of the system:

- ΔH *negative* (exothermic) $\rightarrow$ *T rises*
- ΔH *positive* (endothermic) $\rightarrow$ *T falls*

If the reaction is carried out *isothermally*, the enthalpy change ΔH causes a flow of heat into or from the surroundings, in order to maintain the temperature constant:

- ΔH *negative* (exothermic): $q < 0$ (heat flows *from* system)
- ΔH *positive* (endothermic): $q > 0$ (heat flows *into* system)

Summarizing:

Conditions	*Exothermic* ($\Delta H < 0$)	*Endothermic* ($\Delta H > 0$)
Adiabatic	T rises	T falls
Isothermal	Heat flows from system ($q < 0$)	Heat flows into system ($q > 0$)

3.4 Reaction Enthalpy and Hess's Law

For chemical reactions, we denote the enthalpy change as $\Delta_r H$, known as the reaction enthalpy. When all reactants and products are in their *standard states* (at a pressure $p = p^\ominus = 1$ bar $= 10^5$ Pa), then we refer to the standard reaction enthalpy, $\Delta_r H^\ominus$. It is necessary to specify the temperature at which $\Delta_r H^\ominus$ is defined. This is usually taken to be 25 °C (298 K). To calculate the standard reaction enthalpy for any chemical reaction we use the fact that the enthalpy is a state function, which allows us to conclude (ignoring any mixing effects) that:

"The standard enthalpy of an overall reaction is the sum of the standard enthalpies of the individual reactions into which the reaction may be divided."

Thus, for $a\text{A} + b\text{B} \rightarrow c\text{C} + d\text{D}$:

$$\Delta_r H^\ominus = \left[c\Delta_f H_C^\ominus + d\Delta_f H_D^\ominus\right] - \left[a\Delta_f H_A^\ominus + b\Delta_f H_B^\ominus\right] \quad (3.8)$$

where $\Delta_f H_j^\ominus$ is the standard enthalpy of formation of species j (j = A, B, C or D), and a, b, c, and d are the number of moles of each species involved in the reaction.

In general:

$$\Delta_r H^\ominus = \sum_{\text{prod}} \nu_{\text{prod}} \Delta_f H^\ominus_{\text{prod}} - \sum_{\text{react}} \nu_{\text{react}} \Delta_f H^\ominus_{\text{react}} \quad (3.9)$$

This is known as Hess's Law. The stoichiometric coefficients ν_{prod} and ν_{react} are the smallest integers consistent with the reaction. The $\Delta_f H^\ominus$ values are *defined* to be *zero* for *all elements* in their standard state (p = 1 bar = 100 kPa) at any temperature (real gases are taken to behave ideally in their standard state. For example, $\Delta_f H^\ominus$ for argon is zero even though $H^\ominus(298\text{ K}) = (5/2)RT = 6.2\text{ kJ mol}^{-1}$. Thus the $\Delta_f H^\ominus$ values for all other chemical species are relative to their constituent elements in their standard states. $\Delta_f H^\ominus$ values at 298 K have been determined for many simple compounds, and are tabulated in various textbooks (see Table 3.1 and Further Reading). Note that $\Delta_f H^\ominus$ values are *per mole of the compound formed.* Thus, for the formation of liquid H_2O, the value of $\Delta_f H^\ominus(298\text{ K}) = -285.8\text{ kJ mol}^{-1}$ is for:

$$H_2(g) + 0.5O_2(g) \rightarrow H_2O(l)$$

not for:

$$2H_2(g) + O_2(g) \rightarrow 2H_2O(l)$$

Table 3.1 Examples of enthalpies of formation

Species	$\Delta_f H^\ominus$/kJ mol^{-1}
C(s, graphite)	0
$O_2(g)$	0
$N_2(g)$	0
$H_2O(g)$	−241.8
$H_2O(l)$	−285.8
$CO_2(g)$	−393.5
$H_2O_2(l)$	−187.8
$HN_3(l)$	+264.0
NO(g)	+90.3

Note that $\Delta_f H^\ominus$ can be *negative* or *positive*. We can calculate the standard reaction enthalpy $\Delta_r H^\ominus$ for any reaction, as long as it can be expressed in terms of reactions of species whose standard enthalpies of formation $\Delta_f H^\ominus$ are known (or can be estimated).

Worked Problem 3.2

Q Evaluate the standard reaction enthalpy at 298 K for the following reaction:

$$2HN_3(l) + 2NO(g) \rightarrow H_2O_2(l) + 4N_2(g)$$

A Use the standard enthalpies of formation:

$$2HN_3(l) + 2NO(g) \rightarrow H_2O_2(l) + 4N_2(g)$$

$$\Delta_f H^\ominus \quad 2 \times 264.0 \quad 2 \times 90.3 \quad -187.8 \quad 4 \times 0 \text{ (kJ mol}^{-1}\text{)}$$

$$\Delta_r H^\ominus = [-187.8 + 0] - [528.0 + 180.6]$$

$$\Delta_r H^\ominus \text{ (298 K)} = -896.4 \text{ kJ mol}^{-1}$$

It is important to remember that this is the value *per mole of reaction as written*. Note that enthalpies of combustion, $\Delta_c H^\ominus$, are normally quoted per mole of the species undergoing combustion.

3.5 Temperature Dependence of Enthalpy Changes

Since $C_p = (\partial H/\partial T)_p$ (equation 3.7), it follows that, at constant pressure:

$$d\Delta H = \Delta C_p dT \qquad (3.10)$$

where ΔC_p is the change in heat capacity between the final and initial states. Integrating both sides of this equation from T_1 to T_2 gives:

$$\Delta H(T_2) = \Delta H(T_1) + \int_{T_1}^{T_2} \Delta C_p \mathrm{d}T \qquad (3.11)$$

This important result is known as Kirchhoff's equation. If the C_p are independent of temperature, then:

$$\Delta H(T_2) = \Delta H(T_1) + \Delta C_p(T_2 - T_1) \qquad (3.12)$$

Otherwise we may use, for each species, the empirical form given in equation (3.4):

$$C_p(T) = a + bT + cT^{-2}$$

For reaction enthalpies:

$$\Delta_r H^{\ominus}(T_2) = \Delta_r H(T_1) + \int_{T_1}^{T_2} \Delta_r C_p \mathrm{d}T \qquad (3.13)$$

where:

$$\Delta_r C_p = \sum_{\text{prod}} \nu_{\text{prod}} C_{p,\text{prod}} - \sum_{\text{react}} \nu_{\text{react}} C_{p,\text{react}} \qquad (3.14)$$

If the C_p are independent of temperature, then:

$$\Delta_r H^{\ominus}(T_2) = \Delta_r H^{\ominus}(T_1) + \Delta_r C_p(T_2 - T_1) \qquad (3.15)$$

Summary of Key Points

1. *Heat capacity, C*
   Constant pressure (C_p) and constant volume (C_V) heat capacity; temperature dependence of C_p; phase transitions.

2. *Enthalpy, H*
   Definition, and relationship to C_p; exothermic and endothermic reactions.

3. *Thermochemistry*
   Reaction and formation enthalpy; Hess's law; temperature dependence of reaction enthalpy: Kirchhoff's equation.

Further Reading

E. B. Smith, *Basic Chemical Thermodynamics*, 4th edn., Oxford University Press, Oxford, 1990, chapter 2.
P. W. Atkins, *Physical Chemistry*, 6th edn., Oxford University Press, Oxford, 1998, chapters 2 and 3.
R. A. Alberty and R. J. Silbey, *Physical Chemistry*, 2nd edn., Wiley, New York, 1996, chapter 2.
R. G. Mortimer, *Physical Chemistry*, Benjamin Cummings, Redwood City, Calif., 1993, chapter 2.
D. A. McQuarrie and J. D. Simon, *Molecular Thermodynamics*, University Science Books, Sausalito, Calif., 1999, chapter 5.

Problems

1. Derive expressions for U_m, H_m, C_V and C_p for an ideal gas, and evaluate them at 298 K, using the result from the kinetic theory of gases that $U_m = (3/2)pV$.

2. (i) 100 g of KNO_3 ($\Delta_{sol}H^{\ominus}$(298 K) = +34.9 kJ mol^{-1}) is added to 1 dm^3 of water ($C_p(H_2O)$ = 75.29 J K^{-1} mol^{-1}) at 298 K in an adiabatic container. What is the temperature of the water when the salt has all dissolved?
(ii) Repeat the calculation for $AlCl_3$ ($\Delta_{sol}H^{\ominus}$(298 K) = –329 kJ mol^{-1}).

3. Calculate the reaction enthalpy for the process:

$$0.5N_2(g) + 1.5H_2(g) \rightarrow NH_3(g)$$

at 700 K (the appropriate temperature for ammonia synthesis), given that for NH_3(g), $\Delta_f H^{\ominus}$(298.15 K) = –46.11 kJ mol^{-1}, and the variation of C_p with temperature can be represented approximately by the equation: $C_p = a + bT + cT^{-2}$. Values for the constants a, b and c are:

	a/J K^{-1} mol^{-1}	b/J K^{-2} mol^{-1}	c/J K mol^{-1}
N_2	28.58	3.77×10^{-3}	-0.50×10^5
H_2	27.28	3.26×10^{-3}	$+0.50 \times 10^5$
NH_3	29.75	25.1×10^{-3}	-1.55×10^5

4

The Second and Third Laws: Entropy

We now turn to the topic that lies at the heart of thermodynamics and chemical equilibrium: entropy.

Aims

By the end of this chapter you should be able to

- Understand what determines the spontaneous direction of chemical processes
- Have a qualitative understanding of the term "entropy"
- State the Clausius inequality and understand its implications
- Calculate entropy changes for various processes
- Calculate the temperature and pressure dependence of the entropy

4.1 Spontaneous Processes

What determines the spontaneous direction of change in chemical systems? Some simple examples show that it is not determined solely by the minimization of the energy (or the enthalpy).

(a) Isothermal expansion of an ideal gas. Two flasks having equal volumes are connected by a tap. Initially, flask A is filled with gas to a pressure p_A, and flask B is evacuated. When the tap is opened, gas will flow from A to B until the pressures are equal (Figure 4.1). However, $\Delta U = 0$ and $\Delta H = 0$ for this process [since $(\partial U/\partial V)_T = 0$ and $(\partial H/\partial V)_T = 0$ for an ideal gas; see Chapter 3, Worked Problem 3.1]. Note that heat flows in from the surroundings, because the gas in flask A performs work on the gas in flask B as the pressure p_B builds up from zero to its final value of $p_A/2$.

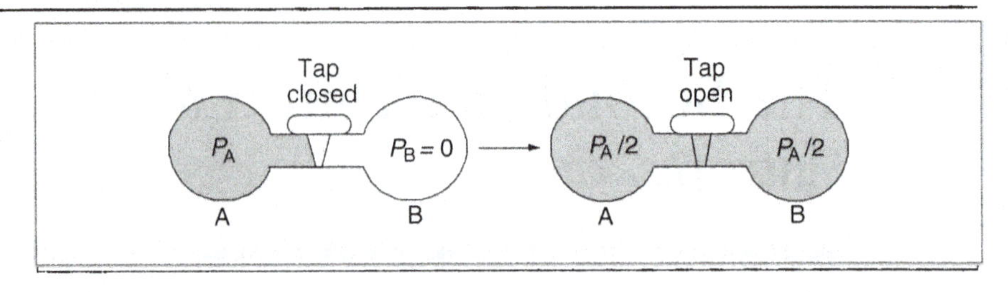

Figure 4.1 Isothermal expansion of a perfect gas

(b) Thermal equilibrium. Two identical blocks of metal, one at a temperature T_A, and the other at a lower temperature T_B, are brought into thermal contact and then isolated (Figure 4.2). Although $\Delta U = 0$ (because they are isolated), we know that heat will flow from A to B until both blocks are at the same final temperature T_f. The chances of the reverse process happening spontaneously are vanishingly small.

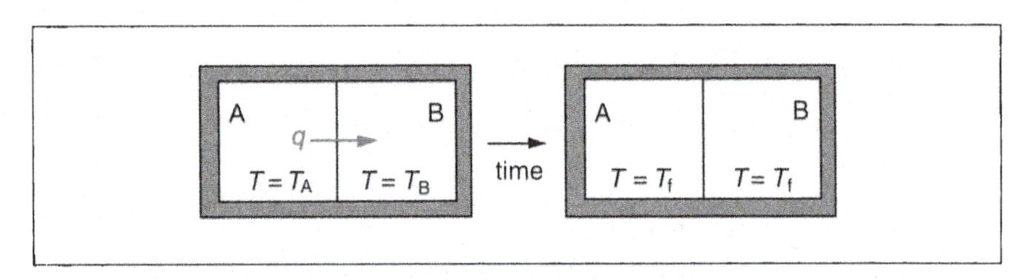

Figure 4.2 Thermal equilibration of two metal blocks, where $T_A > T_B$, and T_f lies between T_A and T_B

(c) Endothermic solution of a salt. If we add a little solid $NaNO_3$ to water, it will dissolve, even though the process is endothermic ($\Delta H > 0$), and hence heat is absorbed (the solution cools down) (Figure 4.3).

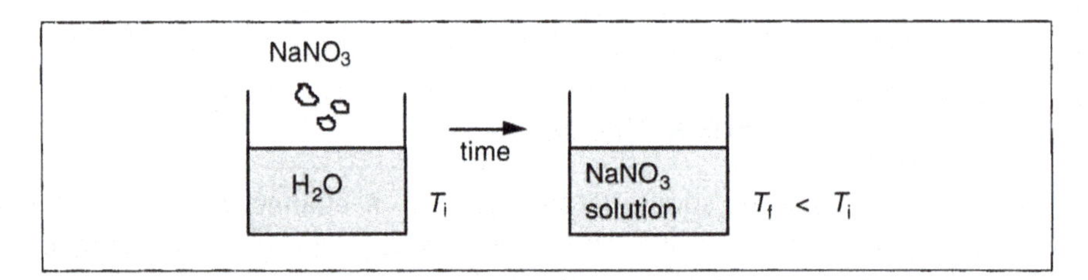

Figure 4.3 Endothermic dissolution of sodium nitrate in water

(d) Chemical equilibrium. Consider the dissociation:

$$N_2O_4 \rightleftharpoons 2NO_2$$

N_2O_4 at 298 K and a pressure of 1 atm will spontaneously partially dissociate (~19%) to NO_2, despite the fact that the reaction enthalpy for dissociation ($\Delta_r H^\ominus = +57.2$ kJ mol^{-1}) is large and positive, *i.e.* the process is endothermic.

The examples (c) and (d) show that reactions can spontaneously occur even if they are *disfavoured* in terms of the enthalpy changes. What then is the driving force? It is clear from these examples that in some way each system is spontaneously moving towards a more disordered or more uniform state. It is also the case (although perhaps less obvious) that the systems have lost some capacity for carrying out work (for example, the expanding gas could have been used to do work on a turbine). We could even have found examples where the system apparently spontaneously becomes more *ordered*, such as the crystallization of a supersaturated solution.

The driving force is that in each case of a *spontaneous process* the total entropy of the system plus its surroundings has increased:

$$\Delta S^{total} = (\Delta S^{system} + \Delta S^{surroundings}) > 0 \qquad (4.1)$$

where ΔS is the change in entropy upon the process occurring. This is a statement of the Second Law of Thermodynamics, which can also be expressed in terms of the small, incremental changes in entropy occurring during the process:

$$\begin{array}{l|l} (\mathrm{d}S + \mathrm{d}S^{surr}) > 0 & \text{spontaneous, irreversible} \\ (\mathrm{d}S + \mathrm{d}S^{surr}) = 0 & \text{equilibrium, reversible} \end{array} \qquad (4.2)$$

Note that for convenience we have dropped the superscript denoting the system, and for the remainder of this text this should be implicitly understood.

The total entropy spontaneously tends to increase until equilibrium is attained, and then stays constant (see Box 4.1).

Box 4.1 Entropy

It is well known that the entropy S of a system is related to the degree of order: a small value of entropy corresponds to a high degree of order (for example, a perfect crystal), whereas a large value of entropy corresponds to a high degree of disorder (for example, a gas). However, this descriptive view of entropy is not always helpful even in qualitatively analysing entropy changes. For example, heating a gas at constant volume leads to an increase in the entropy of the system, but it is perhaps not intuitively obvious that the system has become more disordered. We will see in a later chapter that the entropy is a direct measure of the number of states accessible to the system.

4.2 The Clausius Inequality

Now, example (b) of the previous section suggests that entropy changes must be related in some way to heat flow. When heat flows into the surroundings, it leads to an increase in their disorder. Furthermore, the amount of disorder caused is greater for a given heat flow, the lower the temperature of the surroundings. These observations suggest that the entropy change should be proportional to the heat flow, and inversely proportional to the temperature (later in this text we will justify in more detail why it has this form, and see how to calculate the entropy at a microscopic level). Thus for the *surroundings*:

$$\mathrm{d}S^{\text{surr}} = \frac{\mathrm{d}q^{\text{surr}}}{T^{\text{surr}}} \tag{4.3}$$

Now, if a process occurring in the system causes a heat flow $\mathrm{d}q$ into the system, then the heat flow into the surroundings is $(-\mathrm{d}q)$, and the resulting entropy change is:

$$\mathrm{d}S^{\text{surr}} = \frac{-\mathrm{d}q}{T} \tag{4.4}$$

where we assume that the thermal reservoir of the surroundings is large enough for its temperature to remain the same.

Thus, since $(\mathrm{d}S + \mathrm{d}S^{\text{surr}}) \geq 0$ (from the Second Law):

$$\mathrm{d}S - \frac{\mathrm{d}q}{T} \geq 0$$

or:

$$\mathrm{d}S \geq \frac{\mathrm{d}q}{T} \tag{4.5}$$

This result is known as the Clausius inequality. It states that the change in entropy of the system must be *greater* than $(\mathrm{d}q/T)$ for any *spontaneous*, irreversible process, and will become *equal* to $(\mathrm{d}q/T)$ for a *reversible* process (where equilibrium is maintained at each step). If the system is thermally *isolated*, no heat flow can occur ($\mathrm{d}q = 0$) and so the Clausius inequality becomes:

$$\mathrm{d}S \geq 0 \left| \begin{array}{l} > 0\text{: spontaneous} \\ = 0\text{: equilibrium} \end{array} \right. \tag{4.6}$$

Thus for an isolated system, its entropy will spontaneously increase to some maximum value, which will correspond to equilibrium.

To calculate the entropy S of the system, in order to ensure that S is a state function, we must determine it along a reversible path. Thus we define:

$$\mathrm{d}S = \frac{\mathrm{d}q_{\text{rev}}}{T} \tag{4.7}$$

We can then calculate the entropy change of the system using:

$$\Delta S = \int_{A}^{B} \frac{dq_{rev}}{T} \tag{4.8}$$

Now, although we have to choose a reversible path from A to B in order to be able to calculate the value of ΔS, any other path, in particular any *irreversible* path, will still have this same value, because S is a *state* function.

4.3 Temperature Dependence of the Entropy, S

At constant pressure, the heat flow dq_{rev} into or out of the system to change the temperature from T_1 to T_2 is equal to the incremental enthalpy change dH, which in turn is directly related to the isobaric heat capacity C_p:

$$dq_{rev} = dH = C_p dT \tag{4.9}$$

Thus:

$$\Delta S = S(T_2) - S(T_1) = \int_{T_1}^{T_2} \left(\frac{C_p}{T}\right) dT \tag{4.10}$$

At a phase transition (for example, melting):

$$\Delta_{trs}S = \frac{q_{rev}}{T_{trs}} = \frac{\Delta_{trs}H}{T_{trs}} \tag{4.11}$$

Note that heat flows into the system ($q_{rev} > 0$) to drive an endothermic transition ($\Delta_{trs}H > 0$).

Thus, when a phase transition occurs at some temperature T_{trs}, below the final temperature T_2, and setting $T_1 = 0$ K:

$$S(T_2) = S(0) + \int_{0}^{T_{trs}} \left(\frac{C_p}{T}\right) dT + \frac{\Delta_{trs}H}{T_{trs}} + \int_{T_{trs}}^{T_2} \left(\frac{C_p}{T}\right) dT \tag{4.12}$$

The Third Law states that the entropy of any perfect substance at $T = 0$ may be taken to be $S(0) = 0$. Thus we can determine the absolute (Third Law) value of S at any temperature if we know $C_p(T)$ and $\Delta_{trs}H$. All other transitions are included in the same way (solid–solid, boiling, *etc.*); note that, for second-order transitions, $\Delta_{trs}H = 0$.

Worked Problem 4.1

Q Calculate the standard reaction entropy at 400 K, $\Delta_r S^\ominus(400\ \text{K})$, for the reaction:

$$CO(g) + 2H_2(g) \rightarrow CH_3OH(g)$$

given that the standard Third Law (molar) entropies, $S^\ominus$ at 298 K, and molar heat capacities at constant pressure, C_p, of carbon monoxide gas, hydrogen gas and methanol vapour are:

	$S^\ominus$ /J K^{-1} mol^{-1}	C_p/J K^{-1} mol^{-1}
$CO(g)$	197.7	$28.54 + 2.0 \times 10^{-3}\ T$
$H_2(g)$	130.7	$27.86 + 3.3 \times 10^{-3}\ T$
$CH_3OH(g)$	239.8	$21.75 + 74.9 \times 10^{-3}\ T$

A The standard reaction entropy at 298.15 K is given by:

$$\Delta_r S^\ominus = \sum_{\text{prod}} \nu_{\text{prod}} S^\ominus_{\text{prod}} - \sum_{\text{react}} \nu_{\text{react}} S^\ominus_{\text{react}}$$

Thus:

$$\begin{aligned}\Delta_r S^\ominus(298) &= S^\ominus(CH_3OH, 298) - S^\ominus(CO, 298) - 2S^\ominus(H_2, 298)\\ &= -219.3\ \text{J K}^{-1}\ \text{mol}^{-1}\end{aligned}$$

To find the reaction entropy at a different temperature, T_2, we use equation (4.10) in the form:

$$\Delta_r S(T_2) = \Delta_r S(T_1) + \int_{T_1}^{T_2} \left(\frac{\Delta_r C_p(T)}{T} \right) dT$$

We also have that (equation 3.21):

$$\Delta_r C_p = \sum_{\text{prod}} \nu_{\text{prod}} C_{p,\text{prod}} - \sum_{\text{react}} \nu_{\text{react}} C_{p,\text{react}}$$

Thus:

$$\begin{aligned}\Delta_r C_p &= C_p(CH_3OH) - C_p(CO) - 2C_p(H_2)\\ &= (21.75 - 28.54 - 55.72) + [(74.9 - 2.0 - 6.6) \times 10^{-3}]T\\ &= -62.42 + (66.3 \times 10^{-3})T\end{aligned}$$

The reaction entropy at the higher temperature is thus given by:

$$\Delta_r S^\ominus(400) = \Delta_r S^\ominus(298) + \int_{298}^{400} (-62.42 + 66.3 \times 10^{-3} T) \frac{dT}{T}$$

$$\Delta_r S^{\Theta}(400) = -219.3 - (62.42 \ln T + 66.3 \times 10^{-3} T)$$
$$= -219.3 - 62.42 \ln(400/298) + 66.3 \times 10^{-3}(400 - 298)$$
$$= -219.3 - 18.34 + 6.75$$
$$= -230.9 \text{ J K}^{-1} \text{ mol}^{-1}$$

Worked Problem 4.2

Q Calculate the dependence on temperature and pressure of the entropy of an ideal gas.

A We require an expression for dS, remembering that we must follow a reversible path:

$$\mathrm{d}U = \mathrm{d}q_{\text{rev}} - p\mathrm{d}V$$

Thus:

$$\mathrm{d}q_{\text{rev}} = \mathrm{d}U + p\mathrm{d}V$$

Therefore:

$$= C_V \mathrm{d}T + nRT\left(\frac{\mathrm{d}V}{V}\right)$$

$$\mathrm{d}S = \frac{\mathrm{d}q_{\text{rev}}}{T} = C_V\left(\frac{\mathrm{d}T}{T}\right) + nR\left(\frac{\mathrm{d}V}{V}\right)$$

and:

$$\Delta S = C_V \ln\left(\frac{T_{\mathrm{B}}}{T_{\mathrm{A}}}\right) + nR\ln\left(\frac{V_{\mathrm{B}}}{V_{\mathrm{A}}}\right) \tag{4.13}$$

Under isochoric (V = constant) conditions:

$$\Delta S = C_V \ln\left(\frac{T_{\mathrm{B}}}{T_{\mathrm{A}}}\right) \tag{4.14}$$

Under isobaric (p = constant) conditions, it is straightforward to show that:

$$\Delta S = C_p \ln\left(\frac{T_{\mathrm{B}}}{T_{\mathrm{A}}}\right) \tag{4.15}$$

Under isothermal (T = constant) conditions:

$$\Delta S = nR\ln\left(\frac{V_{\mathrm{B}}}{V_{\mathrm{A}}}\right) \tag{4.16}$$

This last equation can be rearranged to give us the pressure dependence of the entropy (T = constant):

$$S = S^{\ominus} - nR\ln\left(\frac{p}{p^{\ominus}}\right) \tag{4.17}$$

This shows that the entropy of an ideal gas *decreases* with increasing pressure.

Equations (4.14)–(4.17) apply whether the processes are carried out reversibly *or* irreversibly, because S is a state function. However, ΔS^{surr} will in general be different in the two cases. Considering an isothermal expansion of an ideal gas, we know that in the reversible case (equation 2.12):

$$q_{rev} = -w_{rev} = nRT\ln\left(\frac{V_B}{V_A}\right) \tag{4.18}$$

The entropy change of the surroundings is then:

$$\Delta S^{surr} = \frac{-q_{rev}}{T} = -nR\ln\left(\frac{V_B}{V_A}\right) \tag{4.19}$$

and so, for a *reversible expansion*:

$$\Delta S^{total} = (\Delta S + \Delta S^{surr}) = 0 \tag{4.20}$$

However, for the *irreversible* case (*e.g.*, expansion into a vacuum, from V_A to V_B): $q = -w = 0$, and so $\Delta S^{surr} = 0$.

The entropy change of the system, ΔS, has the same value as for the reversible case (because S is a state function), and thus for the spontaneous, irreversible expansion:

$$\Delta S^{total} = \Delta S = nR\ln\left(\frac{V_B}{V_A}\right) > 0 \tag{4.21}$$

This example shows us that the entropy change of the surroundings was needed, in order to tell whether or not the process was spontaneous (the entropy change of the system, ΔS, was the same in both cases). This is not very convenient, as we prefer to focus attention on the *system*. This may be achieved by introducing two new thermodynamic functions, the Gibbs (G) and the Helmholtz (A) free energies, which we will encounter in the next chapter.

Summary of Key Points

1. *Second Law*
   For any spontaneous process, the total entropy (system plus surroundings) must increase. At equilibrium, the total entropy is constant.

2. *Clausius inequality*
   The entropy change of the system must be greater than, or equal to, the heat flow divided by the temperature.

3 *Entropy calculations*
   Entropy changes of the system are calculated along a reversible path. However, since entropy is a state function, the value obtained also applies to any irreversible path.

Further Reading

E. B. Smith, *Basic Chemical Thermodynamics*, 4th edn., Oxford University Press, Oxford, 1990, chapter 3.
P. W. Atkins, *Physical Chemistry*, 6th edn., Oxford University Press, Oxford, 1998, chapters 4 and 5.
R. A. Alberty and R. J. Silbey, *Physical Chemistry*, 2nd edn., Wiley, New York, 1996, chapter 3.
R. G. Mortimer, *Physical Chemistry*, Benjamin Cummings, Redwood City, Calif., 1993, chapter 3.
D. A. McQuarrie and J. D. Simon, *Molecular Thermodynamics*, University Science Books, Sausalito, Calif., 1999, chapters 6 and 7.

Problems

1. It is possible to cool liquids to well below their normal freezing point, because of kinetic barriers to the nucleation of the solid. Suppose 1 mole of supercooled water is stored at –10 °C and 1 bar, but at some point in time it spontaneously freezes to ice.
(i) Why is this an irreversible change?
(ii) Qualitatively discuss how one sets about calculating changes in state functions for irreversible changes.
(iii) Calculate the decrease in entropy $\Delta S^{\ominus}$ for the supercooled water upon freezing, and the resulting increase in entropy of the surroundings, ΔS^{surr}, at the given temperature and pressure. Use the

following: C_p(liq H_2O) = 75.29 J K^{-1} mol^{-1}; C_p(ice) = 36.9 J K^{-1} mol^{-1} (assume these values are independent of temperature within each phase); $\Delta_{fus}H^{\ominus}$(273 K) = +6.008 kJ mol^{-1}.

2. (i) Prove that, under isobaric (constant pressure) conditions, the entropy change ΔS of an ideal gas going from state A to state B is given by equation (4.15):

$$\Delta S = C_p \ln\left(\frac{T_B}{T_A}\right)$$

(ii) Show that, under isothermal conditions, the pressure dependence of the entropy of an ideal gas is given by equation (4.17):

$$S = S \quad - nR\ln\left(\frac{p}{p^{\ominus}}\right)$$

3. Calculate the change in molar entropy, ΔS, of O_2 gas, when it is heated at atmospheric pressure from 25 to 150 °C. The heat capacity is given by:

$$C_p = a + bT + \frac{c}{T^2}$$

The coefficients a, b and c for O_2 gas may be taken to be: a = 29.96 J K^{-1} mol^{-1}; $b = 4.18 \times 10^{-3}$ J K^{-2} mol^{-1}; $c = -1.67 \times 10^5$ J K mol^{-1}.

5

Free Energy

In this chapter we will learn how the entropy changes of the system and surroundings can be incorporated into a new state function of the system, the free energy.

Aims

By the end of this chapter you should be able to

- Define the Gibbs and Helmholtz free energies, and understand their importance
- Establish that the free energy gives the maximum useful work available from the system
- Derive and apply the Fundamental Equation
- Derive the temperature dependence of the free energy (Gibbs–Helmholtz equation)

5.1 Gibbs and Helmholtz Free Energy

The Gibbs and Helmholtz free energies arise from the Clausius inequality, $dS \geq dq/T$, which can be rearranged as $TdS - dq \geq 0$. Since $dq = dU - dw$, we may write:

$$TdS - dU + dw \geq 0 \qquad (5.1)$$

At constant *volume*, $dw = 0$ (assuming no work of any form), giving:

$$(TdS - dU \geq 0)_V \qquad (5.2)$$

At constant *pressure* (assuming no non-pV work), equation (5.1) becomes:

$$TdS - dU - pdV \geq 0$$

or:

$$(TdS - dH \geq 0)_p \tag{5.3}$$

We now define two new functions, the Helmholtz free energy:

$$A = U - TS \tag{5.4}$$

and the Gibbs free energy:

$$G = H - TS \tag{5.5}$$

Differentiating these expressions gives:

$$dA = dU - TdS - SdT$$
$$dG = dH - TdS - SdT$$

If the temperature is held constant ($dT = 0$):

$$dA = dU - TdS$$
$$dG = dH - TdS$$

Thus, substituting into equations (5.2) and (5.3):

$$dA \leq 0; \text{ constant volume} \tag{5.6}$$

$$dG \leq 0; \text{ constant pressure} \tag{5.7}$$

We have thus been able to express the Clausius inequality in terms of two new state functions of the system alone:

	Constant volume	*Constant pressure*
Spontaneous process	$dA < 0$	$dG < 0$
Equilibrium, reversible process	$dA = 0$	$dG = 0$

5.2 Maximum Work

The maximum work is obtained when processes are carried out reversibly. Then the Clausius inequality (equation 4.5) reads: $dS = dq_{rev}/T$, or $dq_{rev} = TdS$. Now we also know from the First Law that $dq_{rev} = dU - dw_{rev}$, and so:

$$dw_{rev} = dU - TdS \tag{5.8}$$

This equation states that:

Maximum work = (Internal energy change) – (Unavailable energy)

Separating the work into pV work plus all "extra" work:

$$dw_{rev}(\text{extra}) - pdV = dU - TdS$$
$$dw_{rev}(\text{extra}) = dU + pdV - TdS$$

Thus the "extra" work under conditions of constant volume and temperature is:

$$dw_{rev}(\text{extra}) = dA \tag{5.9}$$

Under conditions of constant pressure and temperature, the corresponding result is:

$$dw_{rev}(\text{extra}) = dG \tag{5.10}$$

Thus, ΔG is equal to the maximum (non-pV) work available from the system at constant pressure and temperature.

For a spontaneous process under isobaric conditions, $dG < 0$, *i.e.* G will decrease spontaneously until it reaches a minimum (Figure 5.1).

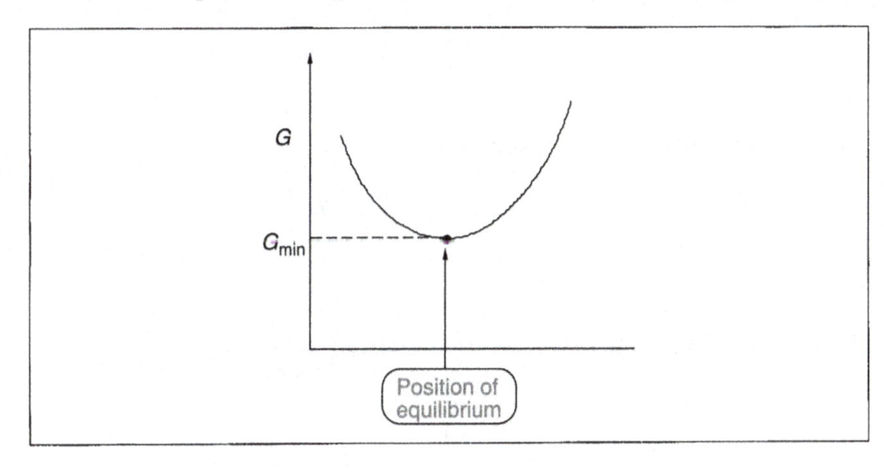

Figure 5.1 The minimum in G gives the equilibrium point

At this point, $dG = 0$, and equilibrium is reached. No further work can be obtained from the system, even under reversible conditions.

5.3 Pressure and Temperature Dependence of G

$$G = H - TS$$
$$= U + pV - TS$$
$$dG = dU + pdV + Vdp - TdS - SdT$$

But $dU = dq + dw$ (First Law), and so $dU = TdS - pdV$ (for a reversible change with no non-pV work). Thus:

$$dG = Vdp - SdT \tag{5.11}$$

This key equation, known as the Fundamental Equation, tells us how G varies with pressure and temperature. Because all the variables are state functions, it applies both to reversible and irreversible processes.

For an isothermal process ($dT = 0$), the Fundamental Equation becomes $dG = Vdp$, and thus gives the pressure dependence of G at constant temperature:

$$\left(\frac{\partial G}{\partial p}\right)_T = V \tag{5.12}$$

For an isobaric process ($dp = 0$), on the other hand, it reduces to $dG = -SdT$, and thus gives the temperature dependence of G at constant pressure:

$$\left(\frac{\partial G}{\partial T}\right)_p = -S \tag{5.13}$$

Now, since $S = (H - G)/T$ (equation 5.5):

$$\left(\frac{\partial G}{\partial T}\right)_p = \left(\frac{G - H}{T}\right)$$

But, from the product rule of differentiation $\left(\frac{d}{dx}(uv) = u\frac{dv}{dx} + v\frac{du}{dx}\right)$:

$$\left(\frac{\partial(G/T)}{\partial T}\right)_p = -\frac{G}{T^2} + \frac{1}{T}\left(\frac{\partial G}{\partial T}\right)_p$$
$$= -\frac{G}{T^2} + \frac{1}{T}\left(\frac{G - H}{T}\right)$$
$$= -\frac{H}{T^2}$$

Thus:

$$\left(\frac{\partial\left(\frac{\Delta G}{T}\right)}{\partial T}\right)_p = -\frac{\Delta H}{T^2} \tag{5.14}$$

This expression is known as the Gibbs–Helmholtz Equation, and tells us how ΔG varies with temperature in terms of ΔH. We will use this result later to calculate the temperature dependence of the equilibrium constant.

Summary of Key Points

1. *Free energy*
   The Gibbs and Helmholtz free energies are state functions of the system, which determine the spontaneous direction of reactions and processes, and equilibrium, under conditions of constant pressure and constant volume, respectively.

2. *Maximum work*
   The free energy gives the maximum useful work which can be done by the system.

3. *Fundamental equation*
   The temperature and pressure dependence of the Gibbs free energy are equal to minus the entropy, and the volume, respectively.

4. *Gibbs–Helmholtz equation*
   The temperature dependence of the Gibbs free energy can usefully be expressed in terms of the enthalpy.

Further Reading

E. B. Smith, *Basic Chemical Thermodynamics*, 4th edn., Oxford University Press, Oxford, 1990, chapter 4.

P. W. Atkins, *Physical Chemistry*, 6th edn., Oxford University Press, Oxford, 1998, chapters 4 and 5.

R. A. Alberty and R. J. Silbey, *Physical Chemistry*, 2nd edn., Wiley, New York, 1996, chapter 4.

R. G. Mortimer, *Physical Chemistry*, Benjamin Cummings, Redwood City, Calif., 1993, chapter 4.

D. A. McQuarrie and J. D. Simon, *Molecular Thermodynamics*, University Science Books, Sausalito, Calif., 1999, chapter 8.

6
Phase Transitions

In this chapter we will explore how the Gibbs free energy determines phase stability for a pure substance, and how the Fundamental Equation can be used to determine the effect of pressure on phase transitions.

Aims

By the end of this chapter you should be able to:

- State what defines the transition point between two phases
- Calculate the effect of pressure on transition temperatures
- Describe the form of pressure–temperature phase diagrams
- Derive and use the Clapeyron and Clausius–Clapeyron equations
- Define the chemical potential, and understand how, for mixtures, the Fundamental Equation must be extended to allow for the effect of changes in composition on the free energy

6.1 Stability of Phases

Under constant pressure conditions, because $\mathrm{d}G < 0$ for a spontaneous process, systems will tend to adopt whichever phase has the lowest value of the Gibbs free energy G. Now $G = H - TS$ (equation 5.5) and $(\partial G/\partial T)_p = -S$ (equation 5.13). Thus G always *falls* with temperature, at a rate given by the entropy S of the phase (Figure 6.1).

Typically, H (solid) < H (liquid) < H (gas) [all *negative*] and S (solid) < S (liquid) < S (gas) [all *positive*]. Thus at low temperatures the solid has the lowest G and is the stable phase. At higher temperatures, first the liquid then the gas become the stable phase. At the phase transition itself, the Gibbs free energies of the two coexisting phases are equal:

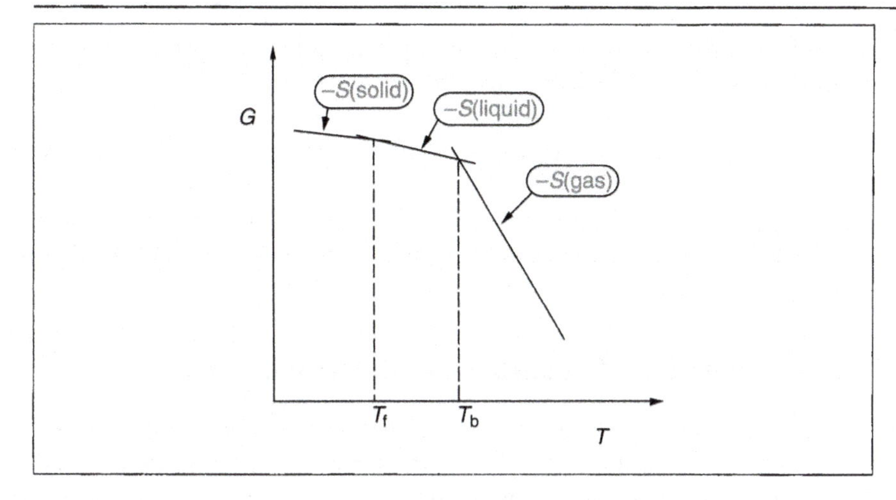

Figure 6.1 The temperature dependence of G for a pure substance

$$\Delta_{trs}G = (G_B - G_A) = 0$$

Therefore, $\Delta_{trs}H = T_t\Delta_{trs}S$, or:

$$\Delta_{trs}S = \frac{\Delta_{trs}H}{T_t} \tag{6.1}$$

Thus we can calculate the transition entropy $\Delta_{trs}S$ directly from the transition enthalpy $\Delta_{trs}H$, which can be measured by calorimetry.

Worked Problem 6.1

Q Calculate the entropy of melting of ice, and the entropy of vaporization of liquid water, given that $\Delta_{fus}H^{\ominus} = +6.008$ kJ mol^{-1} and $\Delta_{vap}H^{\ominus} = +40.656$ kJ mol^{-1}.

A (i) For melting of ice (s) to liquid (l) water, $H_2O(s) \rightarrow H_2O(l)$:

$\Delta_{fus}H^{\ominus} = +6008$ kJ mol^{-1}
$\Delta_{fus}S^{\ominus} = 6008/273.15 = 22.0$ J K^{-1} mol^{-1}
$\Delta_{fus}H^{\ominus}$ is positive (endothermic), and so $\Delta_{fus}S^{\ominus}$ is also positive (*i.e.* the disorder has increased upon melting).

(ii) For vaporization, $H_2O(l) \rightarrow H_2O(g)$:

$\Delta_{vap}H^{\ominus} = +40.656$ kJ mol^{-1} (at 373.15 K). Therefore:
$\Delta_{vap}S^{\ominus} = +109.0$ J K^{-1} mol^{-1}

Note that $\Delta_{vap}S^{\ominus} >> \Delta_{fus}S^{\ominus}$, reflecting the much larger increase in disorder at the boiling point compared to that at the melting point.

It is found that for many simple organic liquids the entropy of vaporization has a constant value close to:

$$\Delta_{vap}S^{\ominus} \approx 85 \text{ J K}^{-1} \text{ mol}^{-1} \tag{6.2}$$

This approximation is known as Trouton's rule. The higher value for water is a reflection of the higher degree of order in liquid water due to hydrogen bonding.

6.2 Effect of Pressure on the Boiling Point

Since $(\partial G_m/\partial p)_T = V_m$, and the molar volume of a gas is much greater than that of the liquid, $V_m(g) >> V_m(l)$, increasing the pressure will slightly increase $G_m(l)$ but greatly increase $G_m(g)$. Thus the crossing points of the curves of $G_m(l)$ and $G_m(g)$, and hence the vaporization temperature T_{vap}, will be shifted to a higher temperature (Figure 6.2).

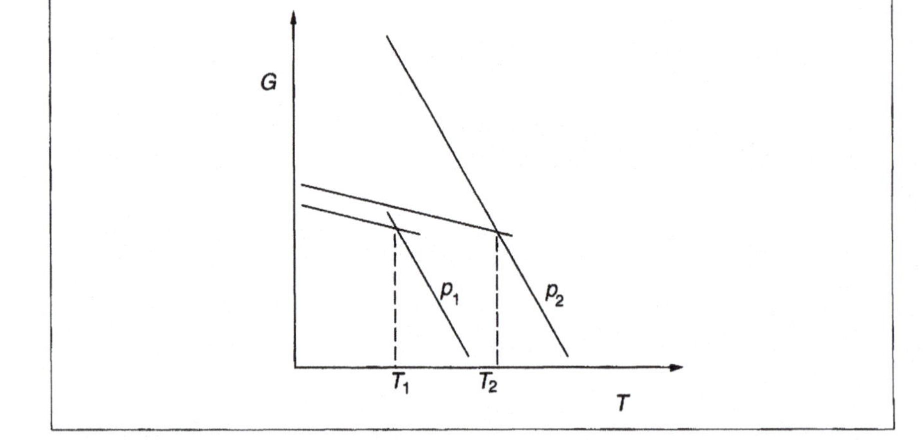

Figure 6.2 The effect of pressure on the Gibbs free energy G for a pure substance ($p_2 > p_1$)

6.3 Phase Diagrams

The variation of the transition temperature with pressure is displayed on a p–T phase diagram (Figure 6.3). Pressure also shifts the melting point T_{fus}, but generally by a smaller amount, since $V_m(s) \approx V_m(l)$. For water, the melting point T_{fus} actually *decreases* with pressure. This unusual behaviour is because ice is less dense than water (it floats!), *i.e.* $V_m(s) > V_m(l)$ for water. Point A is the triple point (T_3 = 273.16 K), where three phases (ice, water and water vapour) coexist.

6.4 Clapeyron Equation

In order to explicitly calculate the two-phase co-existence curves AB, AC and AD (see Figure 6.3), we use the fact that the transition lines are

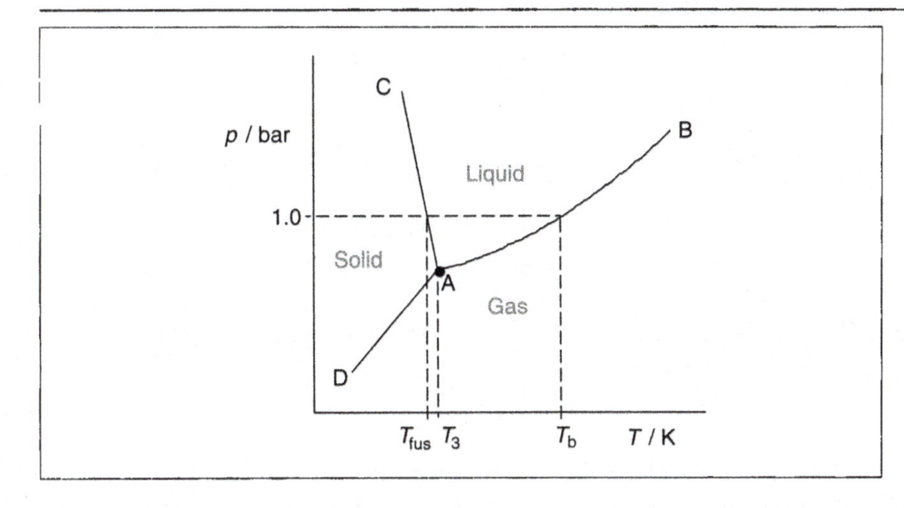

Figure 6.3 Schematic pressure–temperature phase diagram for H_2O (not to scale). T_{fus}, T_3 and T_b are the standard melting point, the triple point and the standard boiling point, respectively

defined by the p and T values where each pair of phases have the same Gibbs free energy G. For example, considering the liquid–vapour equilibrium, if we change p to $(p + dp)$ and T to $(T + dT)$, the two phases will only remain in equilibrium if $dG(l) = dG(g)$; that is, from equation (5.11), if $V(l)dp - S(l)dT = V(g)dp - S(g)dT$.

It follows directly that:

$$\left(\frac{dp}{dT}\right) = \frac{S(g) - S(l)}{V(g) - V(l)} = \frac{\Delta_{vap}S}{\Delta_{vap}V} = \frac{\Delta_{vap}H}{T\Delta_{vap}V} \tag{6.3}$$

The boiling point of a liquid is very sensitive to pressure: for H_2O, an additional pressure of 0.36 atm will increase the boiling point by 10 °C. Similarly for the solid–liquid (melting) curve:

$$\left(\frac{dp}{dT}\right) = \frac{\Delta_{fus}H}{T\Delta_{fus}V} \tag{6.4}$$

The Clapeyron equation is *exact*, and is valid for *all* phase transitions. For H_2O, $\Delta_{fus}V$ is *negative* (-1.7 cm^3 mol^{-1}) and so (dp/dT) is *negative*. Thus ice can be melted by applying pressure. However, the effect is quite small: inserting the value for water of $\Delta_{fus}H^{\ominus} = +\,6.008$ kJ mol^{-1} shows that it needs a pressure of 130 atm to lower the melting point of ice by 1 °C. For most other substances, $\Delta_{fus}V$ is *positive*, and pressure will *increase* the melting point.

6.5 Clausius–Clapeyron Equation

On vaporization, the change in molar volume $\Delta_{vap}V$ may be approximated by $V_m(g)$, since $V_m(g) >> V_m(l)$ by a factor of more than 10^3. Assuming the vapour behaves like an ideal gas, $V_m(g) = (RT/p)$, and substituting into the Clapeyron equation gives:

$$\left(\frac{\mathrm{d}p}{\mathrm{d}T}\right) \approx \frac{p\Delta_{\mathrm{vap}}H}{RT^2}$$

Noting that $\mathrm{d}p/p = \mathrm{d}\ln p$, this may be written as:

$$\left(\frac{\mathrm{d}\ln p}{\mathrm{d}T}\right) \approx \frac{\Delta_{\mathrm{vap}}H}{RT^2} \qquad (6.5)$$

If $\Delta_{\mathrm{vap}}H$ is independent of temperature, this can be integrated to give:

$$\ln p = -\frac{\Delta_{\mathrm{vap}}H}{RT} + \text{constant} \qquad (6.6)$$

Thus, if we measure the vapour pressure p of a liquid as a function of temperature, a plot of $\ln p$ versus $1/T$ will give a straight line of gradient $(-\Delta_{\mathrm{vap}}H/R)$. The explicit expression for the vapour pressure is:

$$p = p^* \exp\left[-\frac{\Delta_{\mathrm{vap}}H}{R}\left(\frac{1}{T} - \frac{1}{T^*}\right)\right] \qquad (6.7)$$

where p^* is the vapour pressure at the initial temperature T^*. The vapour pressure increases steeply as the temperature increases, reaching a value of $p = 1$ atm at the *normal boiling point* T_b. Note that $p = p^{\ominus} = 1$ bar defines the *standard* boiling point (99.6 °C for H_2O), which is thus slightly lower than the boiling point at 1 atm.

Equations (6.5)–(6.7) are equivalent forms of the Clausius–Clapeyron equation.

6.6 Gibbs Free Energy and Chemical Potential

Recall that, for constant temperature, the fundamental equation (5.11) is $\mathrm{d}G = V\mathrm{d}p$. For an ideal gas, $pV = nRT$, and hence:

$$\mathrm{d}G = nRT\left(\frac{\mathrm{d}p}{p}\right)$$

Integrating from p_A to p_B gives:

$$\Delta G = (G_B - G_A) = nRT\ln\left(\frac{p_B}{p_A}\right)$$

If we choose $p_A = p^{\ominus} = 1$ bar, then (writing G_B as G):

$$G = G^{\ominus} + nRT\ln\left(\frac{p}{p^{\ominus}}\right)$$

$$G_m = G_m^{\ominus} + RT\ln\left(\frac{p}{p^{\ominus}}\right)$$

The molar Gibbs free energy $G_m = G/n$ for a pure substance and is equal to the chemical potential μ, which for an ideal gas becomes:

$$\mu = \mu^{\ominus} + RT\ln\left(\frac{p}{p^{\ominus}}\right) \qquad (6.8)$$

where $\mu^{\ominus}$ is the standard chemical potential, *i.e.* the chemical potential at a pressure $p = p^{\ominus} = 1$ bar. This equation shows that we can determine the chemical potential μ(g) for a gas by measuring its pressure, and μ(l) for a liquid by measuring its vapour pressure, since μ(l) = μ(g) when the liquid and vapour are in equilibrium (assuming the vapour behaves ideally).

The formal definition of the chemical potential of a pure substance is:

$$\mu = \left(\frac{\partial G}{\partial n}\right)_{p,T} \qquad (6.9)$$

When we have a *mixture* of different substances, this definition is modified to:

$$\mu_j = \left(\frac{\partial G}{\partial n_j}\right)_{p,T,n_i \neq n_j} \qquad (6.10)$$

where n_j is the number of moles of species j, and the number of moles of all other species present is held constant, along with the pressure and temperature. This equation tells us that the Gibbs free energy will in general depend on the chemical compositions n_j as well as T and p, and so the fundamental equation must be extended (for a binary system) to:

$$\mathrm{d}G = \left(\frac{\partial G}{\partial p}\right)_{T,n_1,n_2}\mathrm{d}p + \left(\frac{\partial G}{\partial T}\right)_{p,n_1,n_2}\mathrm{d}T + \left(\frac{\partial G}{\partial n_1}\right)_{p,T,n_2}\mathrm{d}n_1 + \left(\frac{\partial G}{\partial n_2}\right)_{p,T,n_1}\mathrm{d}n_2 \qquad (6.11)$$

Thus (for a binary mixture):

$$\mathrm{d}G = V\mathrm{d}p - S\mathrm{d}T + \mu_1\mathrm{d}n_1 + \mu_2\mathrm{d}n_2 \qquad (6.12)$$

In general, when a total of N species, labelled j, are present:

$$\mathrm{d}G = V\mathrm{d}P - S\mathrm{d}T + \sum_{j=1}^{N}\mu_j\mathrm{d}n_j \qquad (6.13)$$

This important equation is known as the Fundamental Equation of Chemical Thermodynamics (see Box 6.1).

Box 6.1 Chemical Potential

The chemical potential μ plays an analogous role to the temperature T. Just as heat tends to flow from regions of higher temperature to regions of lower temperature, so matter tends to flow from regions of higher to lower chemical potential.

For the liquid–vapour equilibrium of a pure substance at constant temperature and pressure, equation (6.13) reduces to:

$$\begin{aligned} \mathrm{d}G &= \mu(\mathrm{g})\mathrm{d}n(\mathrm{g}) + \mu(\mathrm{l})\mathrm{d}n(\mathrm{l}) \\ &= [\mu(\mathrm{g}) - \mu(\mathrm{l})]\mathrm{d}n(\mathrm{g}) \end{aligned}$$

If the liquid and vapour phases are in equilibrium with each other, then $\mu(\mathrm{l}) = \mu(\mathrm{g})$ and $\mathrm{d}G = 0$. However, if they are not initially in equilibrium [$\mu(\mathrm{g}) \neq \mu(\mathrm{l})$], then matter will spontaneously transfer from the phase of higher chemical potential to the other phase of lower chemical potential in order to lower the Gibbs free energy ($\mathrm{d}G < 0$). For example, if initially $\mu(\mathrm{l}) < \mu(\mathrm{g})$, then $\mathrm{d}n(\mathrm{g})$ must be negative to ensure that $\mathrm{d}G < 0$. Thus some gas phase molecules will transfer to the liquid phase, until the two chemical potentials become equal.

More generally, for a system with more than one component, the condition for two (or more) phases to be in equilibrium is that the chemical potential μ_i of each component i must be the same within each phase.

Summary of Key Points

1. *Phase stability*
   The most stable phase at constant pressure is the one with the lowest molar Gibbs free energy (chemical potential). For a given phase, its free energy falls with temperature with a (negative) slope equal to the entropy of the phase. Phase transitions are defined by the point where the free energy curves of the two phases cross.

2. *Effect of pressure*
   Pressure increases the free energy, by an amount proportional to the molar volume of the phase; transition temperatures are usually, but not always, increased by pressure.

3. *Phase diagrams*
Pressure–temperature phase diagrams may be calculated using the Clapeyron equation. For ideal gases the Clausius–Clapeyron equation gives the vapour pressure as a function of temperature, which reaches a value of 1 atm at the normal boiling point.

4. *Mixtures*
To describe phase equilibrium for mixtures, the Fundamental Equation must be modified by a number of chemical potential terms (one for each component), to allow for the effect of changes in composition on the free energy of the system.

Further Reading

E. B. Smith, *Basic Chemical Thermodynamics*, 4th edn., Oxford University Press, Oxford, 1990, chapter 4.
P. W. Atkins, *Physical Chemistry*, 6th edn., Oxford University Press, Oxford, 1998, chapters 5–7.
R. A. Alberty and R. J. Silbey, *Physical Chemistry*, 2nd edn., Wiley, New York, 1996, chapters 4 and 6.
R. G. Mortimer, *Physical Chemistry*, Benjamin Cummings, Redwood City, Calif., 1993, chapter 5.
D. A. McQuarrie and J. D. Simon, *Molecular Thermodynamics*, University Science Books, Sausalito, Calif., 1999, chapter 9.

Problems

1. Integrate the Clapeyron equation to derive an approximate expression for the pressure as a function of temperature at the solid–liquid phase boundary. Hence calculate the pressure (in addition to $p^\ominus$) which must be applied to melt ice at –10 °C. Take the molar volumes of ice and liquid water to be 19.7 cm^3 mol^{-1} and 18.0 cm^3 mol^{-1}, respectively, and $\Delta_{fus}H^\ominus$ = +6.008 kJ mol^{-1}.

2. Calculate the boiling point T_b and the enthalpy of vaporization $\Delta_{vap}H$ of butane, given the following data for its equilibrium vapour pressure at a range of temperatures:

T/K	p/mmHg
195.12	9.90
212.68	36.26
226.29	85.59
262.28	503.34
272.82	764.50

3. Calculate the standard Gibbs free energy of formation, $\Delta_f G^{\ominus}$, of liquid H_2O at 298 K. Is the process spontaneous? Analyse the entropy changes occurring in the system and the surroundings, and show how this leads to the same conclusion about the spontaneity of the process.

7
Chemical Equilibrium

In this chapter we will use the concepts and results developed in the preceding chapters to analyse chemical equilibrium and to calculate equilibrium constants.

Aims

By the end of this chapter you should be able to:

- Define the reaction Gibbs free energy
- Understand what defines the position of chemical equilibrium
- Define the relationship between the equilibrium constant and the standard reaction Gibbs free energy
- Derive the temperature dependence of the equilibrium constant
- Understand how pressure affects chemical equilibrium

7.1 Extent of Reaction and the Reaction Gibbs Free Energy

Consider a simple reaction:

$$\mathrm{A} \rightleftharpoons \mathrm{B}$$

For example, the isomerization of pentane to 2-methylbutane. If a small amount dξ of A is converted into B:

$$\mathrm{d}n_{\mathrm{A}} = -\mathrm{d}\xi \quad \text{and} \quad \mathrm{d}n_{\mathrm{B}} = \mathrm{d}\xi$$

The extent of reaction ξ (the Greek letter "xi") is $\xi = 0$ for pure A, and $\xi = 1$ for pure B, for 1 mole of the substance. Now, at constant T and p, if dn_{A} moles of A are converted to dn_{B} moles of B, from equation (6.11):

$$dG = \mu_A dn_A + \mu_B dn_B$$
$$= -\mu_A d\xi + \mu_B d\xi$$

Thus:

$$\left(\frac{\partial G}{\partial \xi}\right)_{p,T} = (\mu_B - \mu_A) \quad (7.1)$$

Note that μ_B and μ_A vary as the extent of reaction ξ varies. This quantity is called the reaction Gibbs free energy, $\Delta_r G$:

$$\Delta_r G = \left(\frac{\partial G}{\partial \xi}\right)_{p,T} \quad (7.2)$$

and it gives the *slope* of the total Gibbs free energy G of the system versus the extent of the reaction ξ (Figure 7.1).

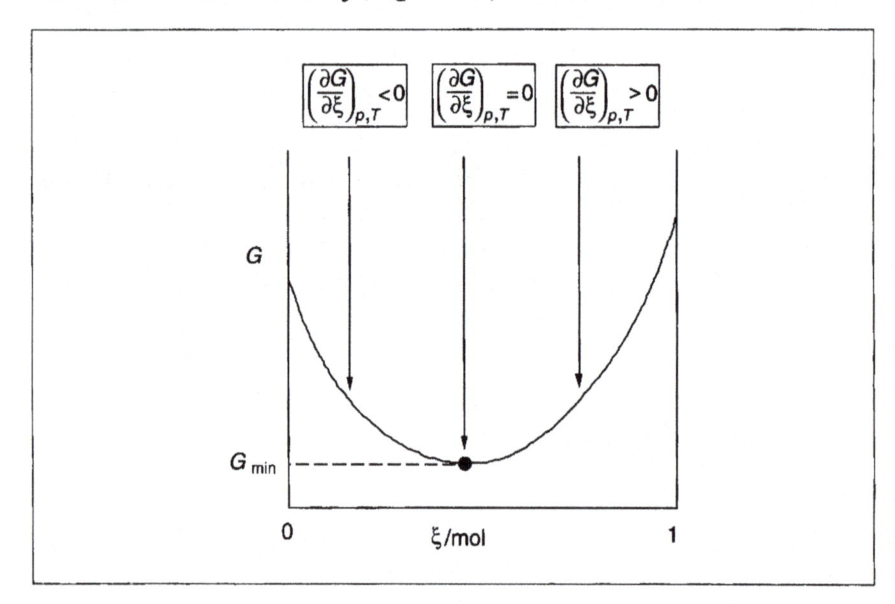

Figure 7.1 Variation of G with the extent of reaction ξ

The symbol $\Delta_r G$, although accepted by IUPAC, is unfortunate because $\Delta_r G$ is in general a derivative of G, not a difference in G. $\Delta_r G$ is only equal to the change in G for 1 mole of A converting to 1 mole of B, at fixed composition (because of the definition of the chemical potentials). This would imply that a large number of moles of both A and B must be present, so that conversion of one mole would hardly change the composition. It is clear that when $\mu_A > \mu_B$, $\Delta_r G$ is negative, G falls with increasing ξ, and the reaction A → B is spontaneous. On the other hand, when $\mu_A < \mu_B$, $\Delta_r G$ is positive, and the opposite reaction B → A is spontaneous. The system will attain equilibrium when $\Delta_r G = 0$, *i.e.* when μ_A

$= \mu_B$ (the chemical potentials of A and B become equal), since this will define the minimum in G.

7.2 The Equilibrium Constant

In order to relate $\Delta_r G$ to the compositions of A and B, we need to substitute expressions for their chemical potentials, μ_A and μ_B. If A and B behave as ideal gases, these are given by:

$$\mu_j = \mu_j^\ominus + RT\ln\left(\frac{p_j}{p^\ominus}\right)$$

Then, the reaction Gibbs free energy $\Delta_r G = (\mu_B - \mu_A)$ is:

$$\Delta_r G = \Delta_r G^\ominus + RT\ln\left(\frac{p_B}{p_A}\right) \tag{7.3}$$

where the standard (molar) reaction Gibbs free energy is:

$$\Delta_r G^\ominus = (\mu_B^\ominus - \mu_A^\ominus) \tag{7.4}$$

Note that $\Delta_r G^\ominus$ *does* represent a free energy difference; it is the standard free energy change when one mole of reaction takes place, with both reactants and products remaining in their standard states ($p = p^\ominus$). We can calculate $\Delta_r G^\ominus$ from the Gibbs free energies of formation $\Delta_f G^\ominus$:

$$\Delta_r G^\ominus = \Delta_f G^\ominus(\mathrm{B}) - \Delta_f G^\ominus(\mathrm{A}) \tag{7.5}$$

Selected values of $\Delta_f G^\ominus$ are listed in various textbooks (*e.g.* see Further Reading). Note that $\Delta_f G^\ominus = 0$ for all elements in their standard states at any temperature [for example, O_2(g), Hg(l), C(s, graphite)].

For the mixture of A and B considered above, if $\xi \approx 0$, $p_A >> p_B$ and so $\Delta_r G < 0$, and the reaction proceeds to the *right*. Conversely, for $\xi \approx 1$, $p_A << p_B$ and so $\Delta_r G > 0$, and the reaction moves to the *left*. Equilibrium occurs when $\Delta_r G = 0$, that is, when:

$$\Delta_r G^\ominus = -RT\ln K_p \tag{7.6}$$

where the equilibrium constant K_p is the ratio of the partial pressures p_A and p_B at equilibrium:

$$K_p = \left(\frac{p_B}{p_A}\right)_{eq} \tag{7.7}$$

If $\Delta_r G^\ominus < 0$, then $K_p = (p_B/p_A)_{eq} > 1$, and equilibrium lies to the *right* (towards B) (Figure 7.2).

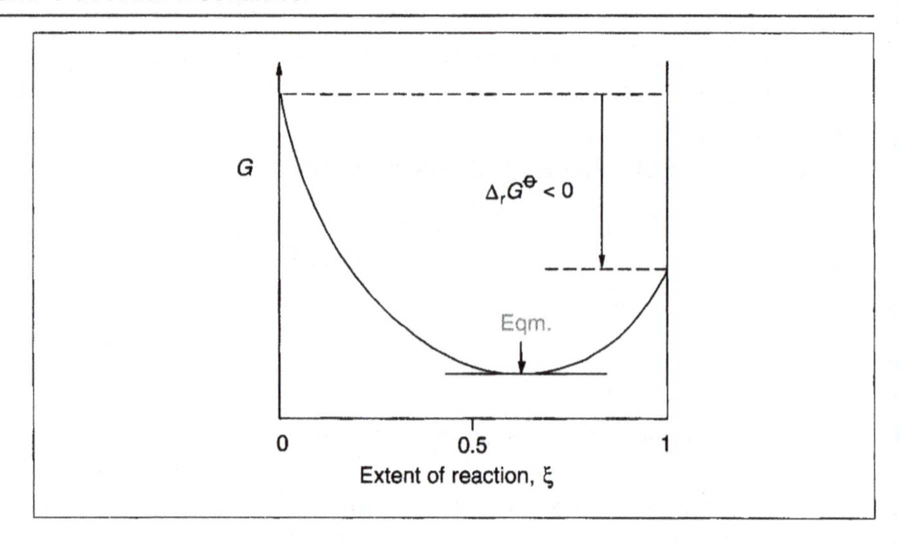

Figure 7.2 Position of equilibrium when $\Delta_r G^\ominus < 0$

If $\Delta_r G^\ominus > 0$, then $K_p = (p_B/p_A)_{eq} < 1$, and equilibrium lies to the *left* (towards A) (Figure 7.3).

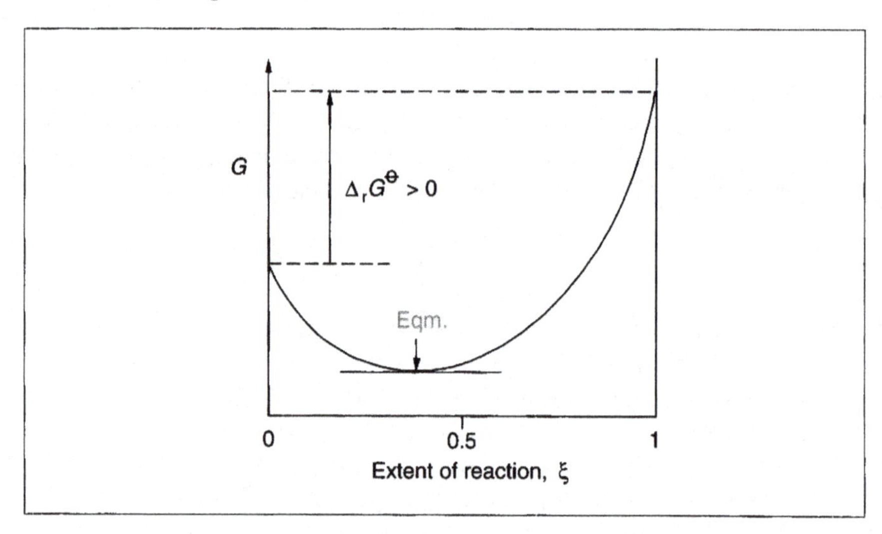

Figure 7.3 Position of equilibrium when $\Delta_r G^\ominus > 0$

If $\Delta_r G^\ominus = 0$, then $K_p = 1$ and $p_B = p_A$. Note that the reason why equilibrium does not lie fully to the left or the right when $\Delta_r G^\ominus \neq 0$ is due to the favourable entropy of mixing of A and B (reactants and products).

For a more complex reaction such as

$$a\text{A} + b\text{B} \rightleftharpoons c\text{C} + d\text{D}$$

we write $\mathrm{d}n_A = -a\mathrm{d}\xi$, $\mathrm{d}n_B = -b\mathrm{d}\xi$, $\mathrm{d}n_C = c\mathrm{d}\xi$ and $\mathrm{d}n_D = d\mathrm{d}\xi$, and hence from equation (6.13) we have:

$$\begin{aligned} dG &= \mu_A dn_A + \mu_B dn_B + \mu_C dn_C + \mu_D dn_C \\ &= -a\mu_A d\xi - b\mu_B d\xi + c\mu_C d\xi + d\mu_D d\xi \end{aligned}$$

and thus:

$$\Delta_r G = (c\mu_C + d\mu_D - a\mu_A - b\mu_B) \tag{7.8}$$

The chemical potentials μ_j are in general expressed in terms of activities a_j, using:

$$\mu_j = \mu_j^\ominus + RT \ln a_j \tag{7.9}$$

The activity is an *effective* concentration, taking into account deviations from ideality due to intermolecular interactions, and takes the following forms (listed here for completeness):

(a) Ideal gas of partial pressure p_j: $a_j = \left(\frac{p_j}{p^\ominus}\right)$;

(b) Real gas of effective pressure (fugacity) f_j: $a_j = \left(\frac{f_j}{p^\ominus}\right)$;

(c) Solvent of mole fraction x_j and activity coefficient γ_j; in this case, $\mu_j^\ominus$ is the chemical potential of pure solvent j: $a_j = \gamma_j x_j$;

(d) Solution of molality (moles of solute per kg of solvent) m_j and activity coefficient γ_j ($\mu_j^\ominus$ is the chemical potential of solute j at unit molality, $m^\ominus = 1$ mol kg^{-1}): $a_j = \gamma_j\left(\frac{m_j}{m^\ominus}\right)$.

In terms of activities the reaction Gibbs free energy becomes:

$$\Delta_r G = \Delta_r G^\ominus + RT\ln\left(\frac{a_C^c\, a_D^d}{a_A^a\, a_B^b}\right) \tag{7.10}$$

where:

$$\Delta_r G^\ominus = (c\mu_C^\ominus + d\mu_D^\ominus - a\mu_A^\ominus - b\mu_B^\ominus) \tag{7.11}$$

Then the equilibrium constant K_a is given by:

$$\Delta_r G^\ominus = -RT \ln K_a \tag{7.12}$$

with:

$$K_a = \left(\frac{a_C^c\, a_D^d}{a_A^a\, a_B^b}\right)_{eq} \tag{7.13}$$

In the most general case, we may write:

$$K_a = \prod_{prod} a_{prod}^{\nu} \prod_{react} a_{react}^{-\nu} \tag{7.14}$$

where $\prod_{\text{prod}}$ and $\prod_{\text{react}}$ denote multiplication over the activities a_{prod} and a_{react} of products and reactants, and the stoichiometric coefficients ν_{prod} and ν_{react} are the smallest integers consistent with the reaction. Note that K_a as defined above is *unitless* (as is K_p).

As before, the values of $\Delta_r G^\ominus$ are obtained from:

$$\Delta_r G^\ominus = \sum_{\text{prod}} \nu_{\text{prod}} \Delta_f G^\ominus_{\text{prod}} - \sum_{\text{react}} \nu_{\text{react}} \Delta_f G^\ominus_{\text{react}} \tag{7.15}$$

(remember that $\Delta_f G^\ominus$ values are tabulated in various textbooks, and are defined to be zero for all elements in their standard state at any temperature). Otherwise, $\Delta_r G^\ominus$ can be obtained from:

$$\Delta_r G^\ominus = \Delta_r H^\ominus - T\Delta_r S^\ominus \tag{7.16}$$

where $\Delta_r H^\ominus$ is obtained from the $\Delta_f H^\ominus$ and:

$$\Delta_r S^\ominus = \sum_{\text{prod}} \nu_{\text{prod}} S^\ominus_{\text{prod}} - \sum_{\text{react}} \nu_{\text{react}} S^\ominus_{\text{react}} \tag{7.17}$$

where the standard Third Law entropies $S^\ominus(T)$ can be obtained by integration of $C_p(T)$ as previously described (Section 4.3), up to the required temperature T.

7.3 Temperature Dependence of the Equilibrium Constant

Since:

$$\ln K = -\frac{\Delta_r G^\ominus}{RT}$$

then:

$$\frac{\mathrm{d}\ln K}{\mathrm{d}T} = -\frac{1}{R}\left[\frac{\mathrm{d}}{\mathrm{d}T}\left(\frac{\Delta_r G^\ominus}{T}\right)\right]$$

(the differentials are complete rather than partial because K and $\Delta_r G^\ominus$ depend only on temperature, not on pressure). Now the Gibbs–Helmholtz equation (equation 5.14) states that:

$$\left[\frac{\partial}{\partial T}\left(\frac{\Delta G}{T}\right)\right]_p = -\frac{\Delta H}{T^2}$$

Hence:

$$\frac{\mathrm{d}\ln K}{\mathrm{d}T} = +\frac{\Delta_r H^\ominus}{RT^2} \tag{7.18}$$

This equation is known as the van't Hoff isochore. If $\Delta_r H^\ominus$ is positive (endothermic reaction), then increasing temperature will increase $\ln K$ and hence K (*i.e.* favour products). If $\Delta_r H^\ominus$ is negative (exothermic reaction), then increasing temperature will decrease $\ln K$ and hence K (*i.e.* favour reactants).

For small temperature changes, we can often assume that $\Delta_r H^\ominus$ is independent of temperature. Integration then gives:

$$\ln\left(\frac{K(T_2)}{K(T_1)}\right) = -\frac{\Delta_r H^\ominus}{R}\left[\frac{1}{T_2} - \frac{1}{T_1}\right] \quad (7.19)$$

Worked Problem 7.1

Q What quantities are required in order to calculate the equilibrium constant K at a temperature T?

A We use: $K = \exp\left[-\frac{\Delta_r G^\ominus(T)}{RT}\right]$

(i) To obtain $\Delta_r G^\ominus(T)$ we either:

(a) use: $\Delta_r G^\ominus = \left(\sum_{\text{prod}} \nu \Delta_f G^\ominus - \sum_{\text{react}} \nu \Delta_f G^\ominus\right)$, looking up the values of $\Delta_f G^\ominus$ at the temperature T, or:

(b) use: $\Delta_r G^\ominus = (\Delta_r H^\ominus - T\Delta_r S^\ominus)$, *i.e.* we need $\Delta_r H^\ominus$ and $\Delta_r S^\ominus$.

(ii) To obtain $\Delta_r H^\ominus(T)$, we use: $\Delta_r H^\ominus = \left(\sum_{\text{prod}} \nu \Delta_f H^\ominus - \sum_{\text{react}} \nu \Delta_f H^\ominus\right)$.

We measure the $\Delta_f H^\ominus$ by combustion ($\Delta_c H^\ominus$), or look them up.

(iii) To obtain $\Delta_r S^\ominus(T)$, we use: $\Delta_r S^\ominus = \left(\sum_{\text{prod}} \nu S^\ominus - \sum_{\text{react}} \nu S^\ominus\right)$.

We either look up the $S^\ominus(T)$ values, or obtain them using:

$S^\ominus(T) = S_0 + \int_0^T \frac{C_p(T)}{T} \mathrm{d}T$, including $\Delta_{trs} S$ for any phase transitions.

(iv) To obtain the $C_p(T)$, we must measure or estimate them.
Thus to calculate K we only need to measure $\Delta_c H^\ominus(T)$ and $C_p(T)$, *and $\Delta_{tr} S$ if any phase transitions occur.*

7.4 Effect of Pressure on Equilibrium

At fixed temperature, K depends only on $\Delta_r G^\ominus$, but this has a value which is defined at the standard pressure of $p^\ominus$ = 1 bar. Thus $\Delta_r G^\ominus$, and hence K, are independent of pressure, that is:

$$\left(\frac{\partial K}{\partial p}\right)_T = 0 \tag{7.20}$$

This does *not* mean, however, that the *compositions* of reactants and products also remain constant. For example, if all reactants and products behave as ideal gases, we can define an equilibrium constant K_x (compare with equation 7.14) in terms of the mole fractions x_j:

$$K_x = \prod (x_{\text{prod}})^{\nu_{\text{prod}}} \prod (x_{\text{react}})^{-\nu_{\text{react}}} \tag{7.21}$$

Thus, since $x_j = p_j/p$:

$$K_x = K_p \left(\frac{p}{p^{\ominus}}\right)^{-\Delta n} \tag{7.22}$$

where $\Delta n = \Sigma\nu_{\text{prod}} - \Sigma\nu_{\text{react}}$. Δn is the change in the total number of moles of gaseous species as the reaction goes from left to right. Thus:

$$\left(\frac{\partial \ln K_x}{\partial p}\right)_T = -\frac{\Delta n}{p} = -\frac{\Delta_r V^{\ominus}}{RT} \tag{7.23}$$

and thus the equilibrium constant in terms of mole fractions, K_x, *does* depend on the pressure. The latter equality is valid for *any* reaction, *i.e.* also in solution. $\Delta_r V^{\ominus}$ is the volume change accompanying one mole of reaction under standard conditions.

7.5 Le Chatelier's Principle

An important empirical principle can be stated as follows:

> *"Perturbation of a system at equilibrium will cause the equilibrium position to change in such a way as to tend to remove the perturbation."*

For example, for an exothermic reaction ($\Delta_r H^{\ominus} < 0$), *lowering* the temperature will shift the equilibrium towards the products, releasing more heat and tending to raise the temperature. Similarly, for a reaction with a positive volume change ($\Delta_r V^{\ominus} > 0$), application of pressure will shift the equilibrium towards the reactants.

Although it is useful to keep in mind these qualitative statements of Le Chatelier's principle, it is also important to put the principle onto a quantitative footing, embodied in the two equations for the effects of temperature (equation 7.18) and pressure (equation 7.23) on chemical equilibrium.

Summary of Key Points

1. *Chemical reactions*
   Extent of reaction. Reaction Gibbs free energy. Position of chemical equilibrium.

2. *Equilibrium constant*
   Relationship between the equilibrium constant and the standard reaction Gibbs free energy. Temperature dependence of the equilibrium constant. Effects of pressure. Le Chatelier's principle.

Further Reading

E. B. Smith, *Basic Chemical Thermodynamics*, 4th edn., Oxford University Press, Oxford, 1990, chapter 4.
P. W. Atkins, *Physical Chemistry*, 6th edn., Oxford University Press, Oxford, 1998, chapter 9.
R. A. Alberty and R. J. Silbey, *Physical Chemistry*, 2nd edn., Wiley, New York, 1996, chapter 5.
R. G. Mortimer, *Physical Chemistry*, Benjamin Cummings, Redwood City, Calif., 1993, chapter 7.
D. A. McQuarrie and J. D. Simon, *Molecular Thermodynamics*, University Science Books, Sausalito, Calif., 1999, chapter 12.

Problems

1. At high temperatures, carbon dioxide dissociates as:

$$2CO_2(g) \rightleftharpoons 2CO(g) + O_2(g)$$

At a pressure $p = 1$ bar, the fractional dissociation α is: $\alpha = 2 \times 10^{-7}$ at 1000 K, and $\alpha = 1.3 \times 10^{-4}$ at 1400 K. Assuming that the standard reaction enthalpy, $\Delta_r H^\ominus$, is independent of temperature, calculate the standard reaction Gibbs free energy, $\Delta_r G^\ominus$, and the standard reaction entropy, $\Delta_r S^\ominus$, at 1000 K.

2. The data in the table below give the equilibrium constant, K_p, as a function of temperature, for the reaction:

$$N_2(g) + O_2(g) \rightleftharpoons 2NO(g)$$

T/K	K_p
1900	2.31×10^{-4}
2000	4.08×10^{-4}
2100	6.86×10^{-4}
2200	1.10×10^{-3}
2300	1.69×10^{-3}
2400	2.51×10^{-3}
2500	3.60×10^{-3}
2600	5.03×10^{-3}

(i) Using these data, calculate the standard reaction enthalpy, $\Delta_r H^\ominus$, and the standard reaction entropy, $\Delta_r S^\ominus$ (assume they are both independent of temperature).
(ii) Using these values, calculate the standard reaction Gibbs free energy, $\Delta_r G^\ominus$, at 1000 K, and hence determine the partial pressure of NO in air at 1000 K, assuming that air consists of 80% N_2(g) and 20% O_2(g).
(iii) The standard Gibbs free energy of formation of NO(g) at 1000 K has a value of $\Delta_f G^\ominus$(1000 K) = +77.77 kJ mol^{-1}. Compare this with the value predicted in part (ii) and discuss any differences.

3. (i) For the gas phase dissociation reaction:

$$A_2(g) \rightleftharpoons 2A(g)$$

derive an expression for the degree of dissociation α in terms of the equilibrium constant K_p.
(ii) For the dissociation of gaseous N_2O_4 to $2NO_2$, calculate K_p and α at pressures p = 1 and 10 atm, for temperatures T = 198, 298 and 398 K, given that $\Delta_r H^\ominus$(298 K) = +57.2 kJ mol^{-1} and $\Delta_r S^\ominus$ (298 K) = +176 J K^{-1} mol^{-1}.

8

The Statistical Definition of Entropy

In the previous chapters of this book we have developed the main concepts of classical thermodynamics, which govern chemical processes at a *macroscopic* level such as we would encounter in the laboratory. However, this begs the question of how this relates to what happens at the *microscopic* or atomic level. The theory that allows us to connect these two extremes is statistical mechanics, and was developed largely towards the end of the 19th century, with the most significant contributions being made by, firstly, Boltzmann and then later by Gibbs. Since much of this work predates the discovery of quantum mechanics, the theory was originally based on a classical or Newtonian view of how atoms moved and interacted with each other. However, given that we have the benefit of this subsequent knowledge, we shall confine ourselves to a quantized view of the world. In other words, energy can only take certain allowed values, known as energy levels, as opposed to be being a continuous function as in classical mechanics.

Statistical mechanics is the means by which we can relate microscopic individual energy levels to macroscopic thermodynamics.

Before we can proceed to develop the ideas of statistical mechanics we must first revisit the concept of entropy.

Aims

In this chapter we will examine the statistical definition of entropy and how this is related to the occupancy of the energy levels of a system. By the end of this chapter you should be able to:

- State the statistical definition of entropy
- Calculate the entropy for simple systems
- Justify the functional form of the statistical entropy

8.1 Statistical Entropy

In Chapter 4 we first encountered the concept of the entropy of a system, and we have already seen that this quantity is a measure of the degree of disorder. Hence we know that a gas has a greater entropy than a liquid phase of the same material, which in turn has a greater entropy than the equivalent solid, because the particles become arranged in a more orderly fashion as we progress along this sequence. Although this allows us to say something qualitatively about what entropy is about, what we need is a more quantitative definition.

It was Boltzmann who first proposed a formal definition of the entropy of a system:

$$S = k_B \ln W \tag{8.1}$$

When **Boltzmann** first proposed the relationship between entropy and microstates it was not widely accepted as being correct. This concerned him and ultimately contributed to his premature demise. As a suitable epitaph, $S = k \ln W$ was engraved on his tombstone. However, we will demonstrate that this hypothesis appears to be borne out in reality.

In this expression, k_B is Boltzmann's constant, which has the value 1.38066×10^{-23} J K^{-1}, ln is the natural logarithm, and W is the number of microstates. We will explain in detail exactly what is meant by the number of microstates in the next section, but this is the term which actually quantifies how disordered a system is.

There are two key points to appreciate concerning the definition of entropy as given in equation (8.1):

- This equation is a *hypothesis*. There is no derivation for this formula: it must be assumed to be true, and its validity rests on whether the results that develop from it are found to be in accord with experimental observation.
- This equation represents the *definition of Boltzmann's constant*. Although Boltzmann's constant and the related universal gas constant, R, appear in many theories and equations in both thermodynamics and kinetics, this is the expression that defines the value and is where it first appeared. We can now begin to appreciate the enormous significance of statistical mechanics, and in particular the statistical definition of the entropy, because it implies that a multitude of other results must all be derivable from this starting point.

$R = N_A k_B$, where N_A is the **Avogadro constant**

Before returning to the subject of the statistical definition of entropy, we must first learn how to calculate the number of microstates for a given system.

8.2 Microstates

Before we can define what a microstate is, it is necessary to understand first what we mean by a state:

> *"The state of a system is defined by specifying the occupancy of all of the energy levels of that system."*

Consider the simple system shown in Figure 8.1, which consists of three evenly spaced energy levels that are occupied by two particles. This system could represent electrons within electronic energy levels or molecules within vibrational energy levels, for instance.

In our simple system we may define the state with the notation (1,1,0) where the numbers, going from left to right, represent the occupancy of the levels from the lowest one upwards. Hence, if we excited one of the particles from the second to third level we would have the distinct state (1,0,1).

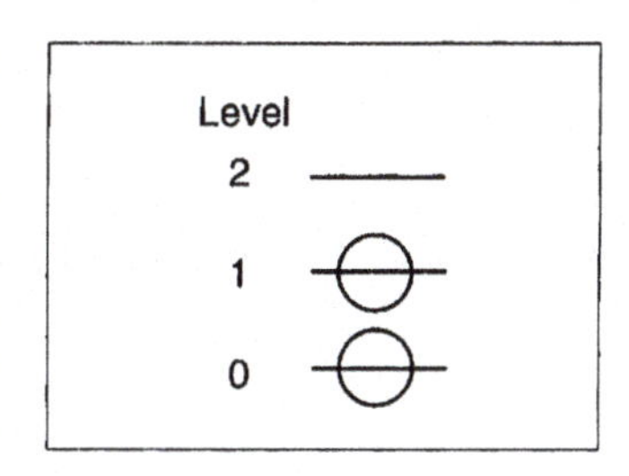

Figure 8.1 A simple three-level system with two particles

We are now ready to address the concept of a microstate. Imagine that in our simple system we can now distinguish between the two particles and we label them 1 and 2. There are now two ways in which we can obtain the state (1,1,0), depending on whether particle 1 or 2 is the one in the lowest level (Figure 8.2). These represent the two possible microstates of the state (1,1,0).

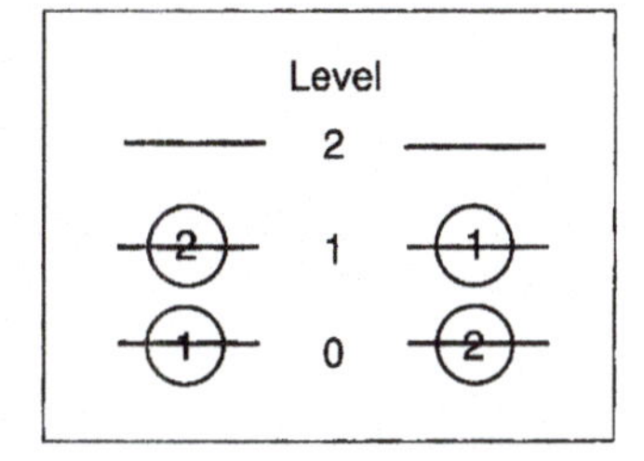

Figure 8.2 The two microstates of the state (1,1,0)

We can formally define a microstate as follows:

"A microstate is a configuration of distinguishable particles within a given state."

The number of microstates, W, is therefore the number of distinct ways of arranging a set of distinguishable particles to achieve a given state.

Worked Problem 8.1

Q For a system consisting of three energy levels with three particles in the state (2,1,0), what is the number of microstates?

A In this case, all we have to do is choose microstates in which each of the three labelled particles in turn is the one on its own in the second level, which leads to three microstates as shown in Figure 8.3.

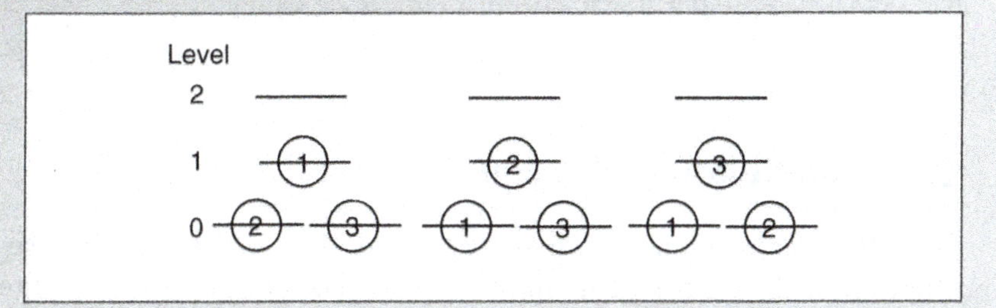

Figure 8.3 The three microstates of the state (2,1,0)

Note that the order of the particles *within* a level is *not* important.

When the number of particles and energy levels is small, then we can derive the number of microstates from inspection by writing down all the possible permutations. However, this rapidly becomes tiresome as the number increases. We therefore need a more general way of arriving at the total number of microstates. Fortunately this is a straightforward matter in probability theory, which leads to the formula:

$$W = \frac{N!}{\prod_{i=1}^{\text{Levels}} n_i!} \qquad (8.2)$$

Product: a sequence of terms that are all multiplied together can be written as follows:

$$\prod_{i=1}^{N} p_i = p_1 \times p_2 \times p_3 \times \ldots \times p_N$$

Sum: a sequence of terms that are all added together can be similarly written as:

$$\sum_{i=1}^{N} s_i = s_1 + s_2 + s_3 + \ldots + s_N$$

Factorials: the value of $n!$ is given by:

$$n! = n \times (n-1) \times (n-2) \times (n-3) \times \ldots \times 2 \times 1$$

An important fact to remember is that $0! = 1$. Consequently, unoccupied levels make no contribution to the number of microstates and can therefore be ignored.

Here N is the total number of particles in the system, whereas n_i is the number of particles in the ith energy level. Therefore the two quantities are related by:

$$N = \sum_{i=1}^{\text{Levels}} n_i \qquad (8.3)$$

Using this general formula for the number of microstates, we can rapidly evaluate this quantity for any given state.

Worked Problem 8.2

Q What is the number of microstates for a system consisting of five energy levels in a state with energy level occupations of (4,3,2,1,0)?

A Using the formula for W, when the total number of particles is 10 and the values of n are as given in the question, we have:

$$W = \frac{10!}{4! \times 3! \times 2! \times 1! \times 0!} = \frac{3628800}{24 \times 6 \times 2 \times 1 \times 1} = 12600$$

8.3 Calculating the Entropy

Now that we have an expression for the number of microstates, the calculation of the entropy for simple systems becomes possible by directly applying the formula proposed by Boltzmann.

Worked Problem 8.3

Q Calculate the entropy for 10 particles in a state with energy level occupations of (4,3,2,1,0).

A This is the system for which we have calculated the number of microstates to be 12600 in the previous problem. Using the formula for S:

$$S = k_B \ln W = 1.38066 \times 10^{-23} \times \ln(12600) \text{ J K}^{-1} = 1.30 \times 10^{-22} \text{ J K}^{-1}$$

Comment: This number is obviously quite small, but it is for only 10 particles. If we were to multiply it by the *Avogadro constant* and divide by 10, we would find the entropy per mole of particles to be approximately 7.8 J K^{-1} mol^{-1}.

We can now begin to justify the definition of statistical entropy, by exploring whether it has some of the basic properties we would expect from what we already know about entropy from classical thermodynamics.

First, let us consider what happens as the temperature tends towards absolute zero. Under these conditions, we know that all particles should be in the ground state energy level, since by definition temperature is a measure of the excess energy in a system above that at 0 K. If all particles are in the same level, then the number of microstates is just equal to *one*.

Hence the *entropy at absolute zero of a system is zero*:

$$S(0 \text{ K}) = k_B \ln 1 = 0 \qquad (8.4)$$

This demonstrates that the statistical definition of entropy is consistent with the Third Law of Thermodynamics.

Secondly, let us consider what happens if we treble the size of our original simple system composed of two particles in the state (1,1,0) so that we have three sets of levels and three pairs of particles (Figure 8.4).

Note that we treat the three individual copies within the trebled system as being physically separated so that the particles cannot be exchanged. This yields a total of eight microstates overall. Let us now consider the entropy of the original two-particle system, and of our new system, which is three times as large:

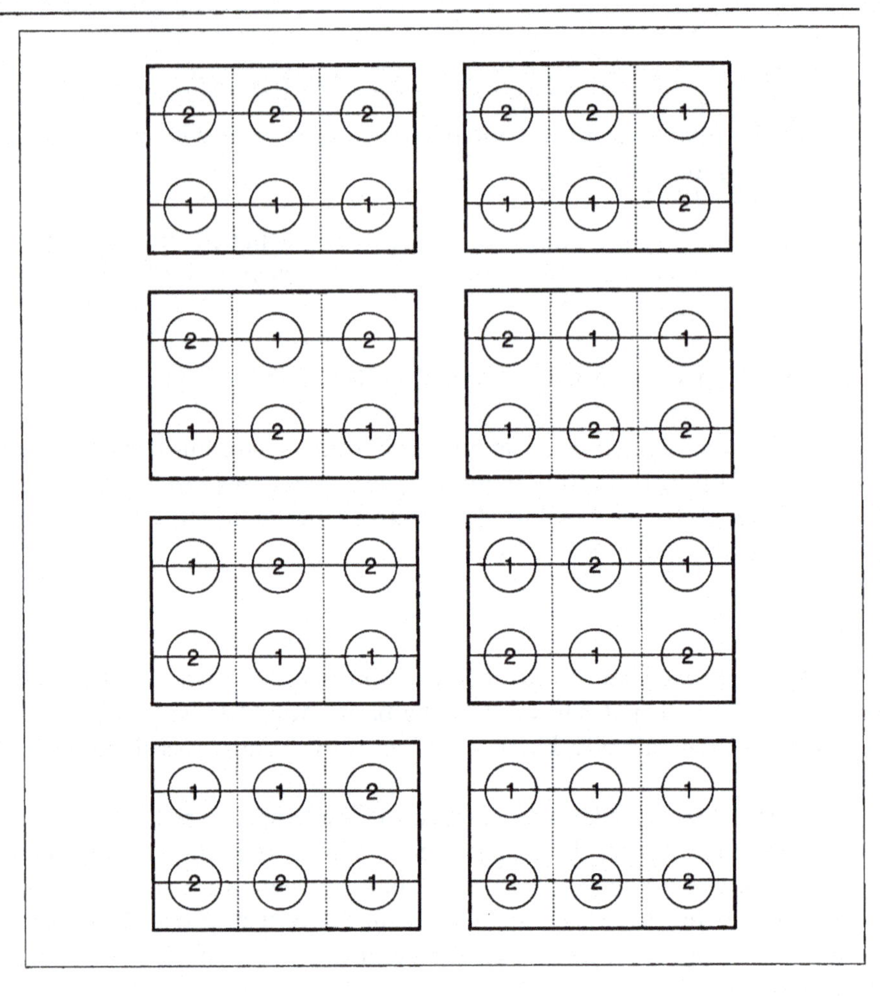

Figure 8.4 The eight microstates for the trebled system size of the state (1,1,0)

For two particles:

$$S_2 = k_B \ln 2 \tag{8.5}$$

For six particles:

$$S_6 = k_B \ln 8 = k_B \ln 2^3 = 3k_B \ln 2 \tag{8.6}$$

From the above example we can see that when the system size is trebled, then the statistical entropy also trebles. Thus, as required, the entropy is an extensive property of the system.

To express things more generally, the number of microstates increases as 2^N, where N is the number of copies of the original system. Because the number of microstates therefore increases exponentially with the size of the system, it follows that the entropy must be proportional to the natural logarithm of the number of microstates, as hypothesized by Boltzmann. Hence, although we cannot actually prove the statistical definition of entropy, it can be demonstrated to be consistent with the known properties of classical entropy.

Summary of Key Points

1. *The statistical definition of entropy*
   The entropy of a state is proportional to the natural logarithm of the number of microstates within that state.

2. *Boltzmann's constant*
   The constant of proportionality between statistical entropy and the natural logarithm of the number of microstates is Boltzmann's constant, k_B. This is the definition of Boltzmann's constant, and all other occurrences of k_B can be derived from here.

3. *Microstates*
   The number of microstates, and thus the statistical enrtopy, can be calculated if the occupancy of the energy levels of the system in a given state is known. In order to determine the number of microstates, it is necessary to calculate how many ways a set of distinguishable particles can be arranged over the energy levels in accordance with the desired occupancies.

Further Reading

J. S. Dugdale, *Entropy and its Physical Meaning*, Taylor and Francis, London, 1996.

R. P. H. Gasser and W.G. Richards, *Entropy and Energy Levels*, Clarendon Press, Oxford, 1974.

R. K. Pathria, *Statistical Mechanics*, 2nd edn., Butterworth-Heinemann, Oxford, 1996.

D. H. Trevena, *Statistical Mechanics: an Introduction*, Ellis Horwood, Chichester, 1993.

Problems

1. Experimental estimates of the absolute entropy of solid carbon monoxide close to absolute zero suggest that $S(0\ K) \approx 5.7\ J\ K^{-1}\ mol^{-1}$, not zero as expected. Suggest a reason why this might be the case.

2. If four particles were placed in an infinite set of uniformly spaced energy levels, how would they occupy the energy levels in order to maximize the entropy?

3. Calculate the entropy for a system consisting of 10 particles distributed over four energy levels with occupancies of (5,3,2,0).

4. If a single particle were to be excited by one energy level in the system of problem 3, what would be the occupancy of the levels that maximizes the entropy of the system? If the separation of the energy levels is 1×10^{-20} J, what would be the Helmholtz free energy change for this process at 298 K? [*Note*: Helmholtz free energy differences can be described using an equivalent expression to equation (7.16) for the Gibbs free energy, $\Delta A = \Delta U - T\Delta S$]

9 Connecting Microscopic and Macroscopic Properties

So far we have shown how statistical mechanics allows the calculation of the entropy of a system from the occupation of the energy levels. Initially, it may seem that the present chapter departs somewhat from this theme, as we consider how it is possible to relate the macroscopic properties of a system to the microscopic states and energy levels. However, towards the end we will see that the statistical definition of the entropy actually plays a crucial role in quantifying this connection.

Aims

In this chapter we will introduce the concept of an ensemble, and demonstrate that the macroscopic properties of a system can be represented as an ensemble average. By introducing the idea of the probability of a state being related to its free energy, we will develop an expression for the distribution of particles over energy levels. By the end of this chapter you should be able to:

- State the definition of an ensemble and give an example
- Express the macroscopic properties of a system in terms of the values for each state
- Calculate the most probable distribution of particles over the energy levels

9.1 Ensembles

At the macroscopic level that we encounter in the laboratory, the state of a system can be defined by invoking a range of quantities, including those given in Table 9.1.

Table 9.1 Examples of different types of macroscopic parameters

Parameter	*Symbol*	*Comments*
Number of atoms	N	
Number of moles	n	$N = nN_A$
Volume	V	
Density	ρ	Related to N or n and V
Internal energy	U	Sum of kinetic and potential energy contributions
Temperature	T	Measure of kinetic energy
Pressure	p	Force exerted by a gas per unit area due to kinetic energy

Although there are a large number of different macroscopic variables, we can see that many of them are related and therefore not independently adjustable. For example, we have already encountered the ideal gas equation:

$$pV = nRT \tag{9.1}$$

In this case we can clearly see that once we know, for instance, the number of moles of gas, the volume they occupy and the temperature, then the pressure is determined.

A **phase** is a state of matter that is generally uniform through out, both in physical properties and chemical composition. Most chemical substances exhibit at least three phases, namely *solid*, *liquid* and *gas*. However, many systems can exist in more phases than this. For example, a solid can have several distinct structures with the same composition known as *polymorphs*, each one of which represents a distinct phase.

It turns out that the above idea that only *three* macroscopic variables of a system can be independently controlled is quite general for a *single phase*. When an experiment is performed, the set of three variables that is effectively fixed or controlled depends on the way in which the experiment is carried out. A particular set of three fixed variables is called an ensemble, some examples of which are given in Table 9.2.

Table 9.2 Examples of common ensembles

Ensemble name	*Fixed variables*	*Laboratory conditions*
Microcanonical	N, V, U	Isolated system in a vacuum
Canonical	N, V, T	Enclosed system in a heat bath
Isothermal/isobaric	N, p, T	Open system in a heat bath
Grand canonical	μ, V, T	Temperature-controlled adsorbent in contact with a gas

We should note that, although in Table 9.2 the microcanonical ensemble has the fixed variables N, V and U according to the current nomenclature, this is more generally referred to as the NVE ensemble

Perhaps the most widely employed experimental conditions are those in which a sample is thermostated in an open vessel, and therefore the isothermal/isobaric ensemble is the most important. For simplicity, most of the arguments we will develop in this book will actually relate to the canonical ensemble. However, the same principles can be applied to other sets of conditions or ensembles as well.

9.2 Ensemble Averages

What we have said about ensembles so far just defines the types that exist and the basic concept of having fixed macroscopic quantities. Now we must give a more formal definition, from which it will become apparent where the name "ensemble" actually comes from:

> *"An ensemble is a set of sampled configurations of a system, in contact with a thermal bath, where each individual sample may have a different state while sharing three common macroscopic properties."*

At first sight this sounds a bit daunting; however, it should become clearer with an example. The significance and utility of this construction will become more apparent in the next section of the chapter. Imagine we have a system that consists of a diatomic molecule in a box of fixed volume, which is only undergoing simple harmonic vibration. Now let us draw many sampled configurations of this system and place them together, as schematically illustrated in Figure 9.1.

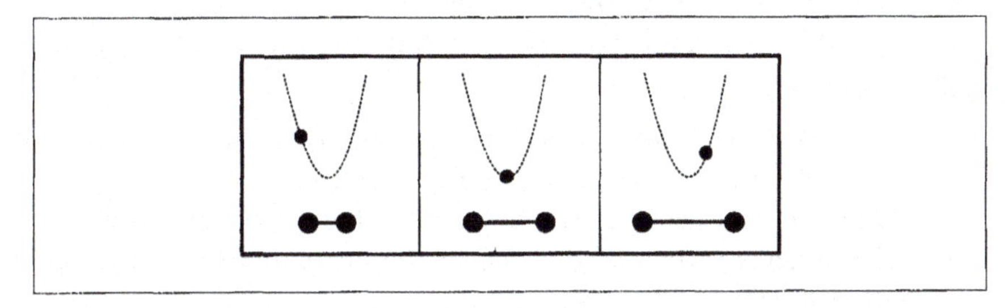

Figure 9.1 Three, from a series of many, sampled configurations of a system consisting of a single diatomic molecule. *Inset* is the potential energy curve for simple harmonic motion, indicating the stage each molecule has reached

Providing the size of the box is sufficiently large for the molecules in adjacent boxes not to interact, there will be no exchange of energy and the volume will remain constant. Hence we have a situation where the system is in the *microcanonical ensemble*.

During simple harmonic motion there will be a continual interchange between potential and kinetic energy within each molecule, as shown in Figure 9.2.

Turning our attention to the properties of this particular ensemble that are not constrained to be the same in different sampled configurations, let us consider the temperature within each box. Temperature in this case is simply a measure of the kinetic energy, and so will be varying with time in any given box. However, the average value within any

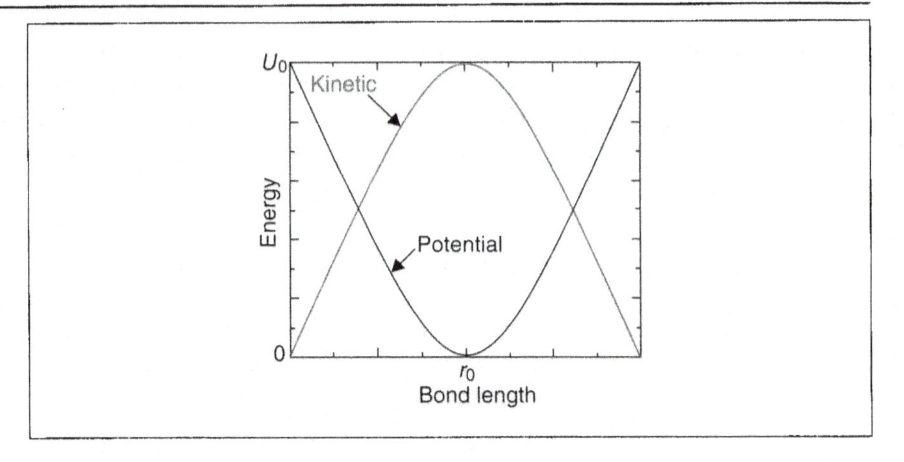

Figure 9.2 Schematic illustration of the potential and kinetic energy versus bond length for a harmonic oscillator. Here U_0 is the initial energy imparted to the system

given box will be constant, since all molecules are undergoing the same harmonic motion.

If we assume that different sampled configurations of our diatomic molecule started vibrating at random times, then each sample may have a distinct amount of kinetic energy and therefore a different instantaneous temperature. The average temperature of all the sampled configurations at any instant will tend to be close to the average temperature of any individual sampled configuration as it evolves over time, and the degree of fluctuation about the average will be small. How true this is obviously depends on just how out of phase all the vibrations are within the evolving sampled configurations.

From this "thought experiment" we can draw a number of important general points:

- All properties, which are *not* specified as fixed within a given ensemble, *fluctuate* about an average value.
- When the average value is *constant*, then the system is at equilibrium.
- The larger a system is, the smaller the fluctuations about the average will be.
- For a sufficiently large number of sampled configurations, the average value of a property over the ensemble is the same as the time average of that property for a single system.

This last point, which states that the ensemble average, $<P>_{\text{ensemble}}$, and the time average should be the same, is known as the ergodic hypothesis

Worked Problem 9.1

Q Think of a system for which the ergodic hypothesis fails.

A The ergodic hypothesis makes no allowance for kinetics, and therefore any potential energy surface with two, or more, local minima with barriers between them could lead to a failure. Consider the energy surface in Figure 9.3, with two minima, A and B.

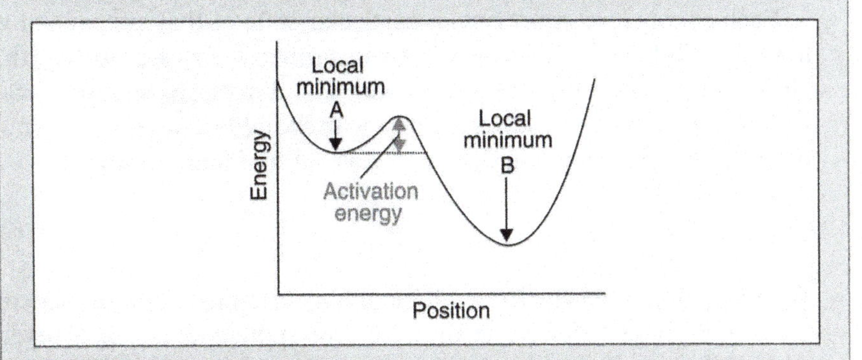

Figure 9.3 A potential energy surface with two local minima

If a single sample configuration is started in one minimum, for instance A, and the kinetic energy is insufficient to overcome the barrier to reach B, then the time average properties will reflect only one part of the potential energy surface. However, the ensemble average may contain information about both minima if the different sampled configurations were started with random positions along the potential curve.

To put things on a more formal footing, we can express the value of some macroscopic measurable property, P, according to the following:

The **ensemble average** of a property P can be written as:

$\langle P \rangle_{\text{ensemble}}$

This uses the notation that ⟨ ⟩ indicates the average of the quantity within the brackets. Formally, this implies the integral over all possible configurations. However, if the number of configurations is finite, then this becomes a summation.

$$P_{\text{macro}} = \langle P \rangle_{\text{ensemble}} = \frac{1}{N_e}\left(\sum_{i=1}^{N_e} P_i\right) \qquad (9.2)$$

where N_e is the number of sampled configurations in the ensemble and P_i is the value of the property for the ith configuration.

Considering the ensemble average as being over a large number of copies of the system that sample different configurations is not particularly convenient in practice, since this is a hypothetical construct. The reality is that we have one system that explores all the states that are consistent with the ensemble being considered. Hence we can rewrite equation (9.2) in terms of these N_s states and the probability, ρ_i, of the ensemble being in that state:

$$P_{\text{macro}} = \left(\sum_{i=1}^{N_s} \rho_i P_i\right) \qquad (9.3)$$

Another way of viewing this change is to take our large number of sampled configurations within the ensemble and to classify them all according to what state they are in. The probability of a state then just becomes the number of sampled configurations in that particular state, divided by the total number of samples.

Each energy level represents a particular state of that system and will have a well-defined value of any given property, P, associated with it. Hence, in equation (9.3) we can equally as well replace the sum of states with a sum over energy levels where the probability is now just the number of particles occupying that level divided by the total number:

$$\rho_i = \frac{n_i}{N} \tag{9.4}$$

We have almost reached our ultimate aim of relating the macroscopic value of a property for a system to that for an individual energy level or state. All that is needed now is to know how the particles will be distributed over the available energy levels.

9.3 What is the Preferred State of a System?

To understand how we go about determining the most favourable distribution of a given number of particles over a set of energy levels, it is easiest to consider a simple example.

Worked Problem 9.2

Q A system consists of a set of uniformly spaced energy levels of separation 1×10^{-20} J, starting from a first level which has an energy of 0 J. If six particles are placed within this system with a total energy of 4×10^{-20} J, how many possible states are there that satisfy these requirements?

A To answer the question we just need to draw a set of energy levels and carefully think of all the permutations of six particles that have the correct total energy when summed up (Figure 9.4).

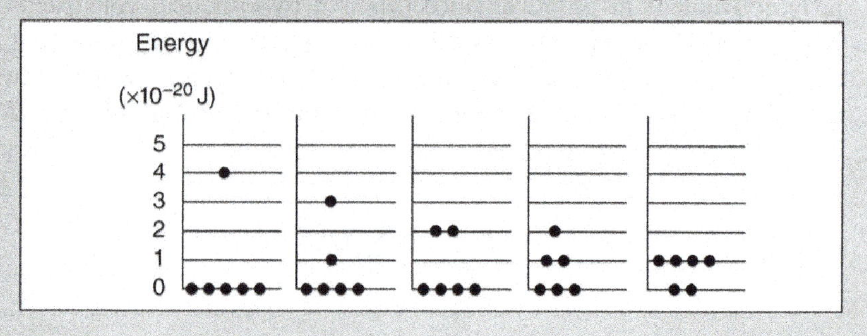

Figure 9.4 All possible states that satisfy the criteria of the question

The answer is that there are five valid distributions of the particles over the energy levels.

Now let us consider the outcome of the above problem more closely. Although there are five different distributions that have the correct energy and number of particles, we now have to consider the question of whether each of these distributions is equally probable.

From our knowledge of classical thermodynamics, we expect that a system will always try to *minimize* its *free energy*. In the case of this problem we are implicitly working under conditions of fixed volume, since in general if we were to change the volume this would alter the energy level spacing. Consequently the relevant quantity is the *Helmholtz free energy*, *A* (see Section 5.1).

Worked Problem 9.3

Q For the system in the previous worked problem, which of the possible distributions has the lowest Helmholtz free energy at room temperature (298.15 K)?

A We can now use the results from the previous chapter (the ones which tell us how to calculate the entropy based on the number of particles in each level, *i.e.* equations 8.1/8.2) for the five different distributions:

$$W = \frac{6!}{n_1!n_2!n_3!n_4!} \tag{9.5}$$

$$S = k_B \ln W \tag{9.6}$$

$$A = U - TS \tag{9.7}$$

State	*W*	*S* ($\times 10^{23}$ J K^{-1})	*U* ($\times 10^{20}$ J)	*A* ($\times 10^{20}$ J)
1	6	2.47	4.00	3.26
2	30	4.70	4.00	2.60
3	15	3.74	4.00	2.89
4	60	5.65	4.00	2.32
5	15	3.74	4.00	2.89

The state with the lowest free energy is number 4.

Comment: Because all of the distributions have the same internal energy by definition, the answer will not depend on the temperature specified.

For small numbers of particles it is quite feasible to follow the above approach to finding the preferred distribution that minimises the free energy. However, for real macroscopic systems we have to be able to deal with numbers of particles of the order of Avogadro's constant. Fortunately, Boltzmann has already generalized this method for the limit of a large number of particles to yield the distribution that carries his name.

9.4 The Boltzmann Distribution

The distribution of particles over a series of energy levels can be found by following the same procedure as above, but in an algebraic fashion. Hence the basic principles are:

- *Maximize the entropy* of the system
- Constrain the *number of particles, N,* to be *constant*
- Constrain the *internal energy, U,* to be *constant*

The derivation of the Boltzmann distribution is beyond the scope of this text, though this can be found in the Further Reading for those who are interested. The most important thing is to know the final result, and to understand the principles from which it is derived:

$$n_i \propto \exp\left(-\frac{U_i}{k_B T}\right) \tag{9.8}$$

$$\rho_i = \frac{n_i}{N} = \frac{\exp\left(-\frac{U_i}{k_B T}\right)}{\sum_{i=1}^{\text{all levels}} \exp\left(-\frac{U_i}{k_B T}\right)} \tag{9.9}$$

There is one further key point to understand: the Boltzmann distribution is an *approximate result* that becomes valid in the limit of *large numbers of particles*. Therefore, it would have been wrong to use this result to tackle the worked problems given earlier, but it is perfectly valid when we are dealing with, for example, a mole of particles.

The distribution of particles over energy levels according to the Boltzmann distribution is illustrated schematically in Figure 9.5. This demonstrates that, as the temperature tends to absolute zero, all particles should be in the ground state, while as the temperature becomes very large relative to the energy separation, then the difference in occupancy between adjacent states decreases.

Now that we have an expression for the probability of finding a particle in an energy level for a macroscopic system, we are in a position to be able to calculate the properties of that system starting from those of individual atoms or molecules. Exactly how this can be done is the subject of the next chapter.

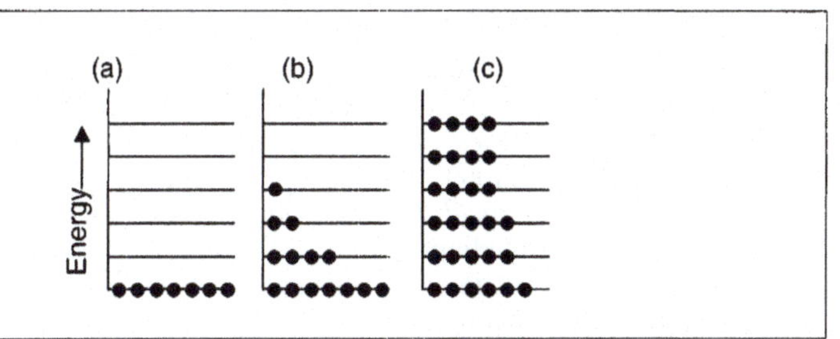

Figure 9.5 Schematic illustration of the occupancy of a set of energy levels of separation Δu when (a) $T = 0$, (b) $k_BT \approx \Delta u$, (c) $k_BT >> \Delta u$

Summary of Key Points

1. *Ensembles*
   For any single-phase system, three macroscopic state variables (*e.g.* N, V, U, T) may be fixed to give rise to a particular ensemble. The remaining variables fluctuate about a steady value when the system is at equilibrium.

2. *Ensemble averages*
   The macroscopic average of a property of an ensemble is equivalent to the sum over the microscopic values for particular energy levels or states, weighted by the probability of the level being occupied.

3. *The ergodic hypothesis*
   This hypothesis proposes that the average of a property over many sampled configurations of a basic component of a macroscopic system is the same as the time average of that property for any single part. However, this is only true when there are no barriers that prevent all the possible states from being explored.

4. *The preferred state of a system*
   The most favourable state of a system is the one that maximizes the statistical entropy, subject to the constraints of fixed numbers of particles and fixed total energy.

5. *The Boltzmann distribution*
   The probability of a particle occupying a given energy level is proportional to the exponential of the negative of its internal energy divided by Boltzmann's constant and the absolute temperature (in K):

$$n_i \propto \exp\left(-\frac{U_i}{k_B T}\right)$$

Further Reading

P. W. Atkins, *Physical Chemistry*, 6th edn., Oxford University Press, Oxford, 1998, chapter 19.
C. E. Hecht, *Statistical Thermodynamics and Kinetic Theory*, Freeman, New York, 1990, chapter 1.
R. K. Pathria, *Statistical Mechanics*, 2nd edn., Butterworth-Heinemann, Oxford, 1996, chapters 1–3.
D. H. Trevena, *Statistical Mechanics: an Introduction*, Horwood, Chichester, 1993, chapter 3.

Problems

1. A system consists of a set of uniformly spaced energy levels of separation 1×10^{-20} J, starting from a first level which has an energy of 0 J. Eight particles are placed within this system with a total energy of 6×10^{-20} J. (a) How many possible states are there that satisfy these requirements? (b) Which state is the most probable?

2. If there exists two excited states at energies of 0.72 and 1.24 kJ mol^{-1} above the ground state in a system, what would be the percentage of particles occupying each state at equilibrium when the temperature is 300 K?

10
The Partition Function

In the previous chapter we demonstrated that some property, P, can be related to the value for an individual energy level, P_i, according to the probabilities, ρ_i, given by the Boltzmann distribution:

$$P_{\text{measured}} = \langle P \rangle_{\text{ensemble}} = \sum_{i=1}^{\text{all levels}} \rho_i P_i \tag{10.1}$$

Now we will develop the above approach further through the use of a quantity called the partition function

Aims

In this chapter we will define the partition function, and show how the key macroscopic thermodynamic properties can be written in terms of this single quantity. By the end of this chapter you should be able to give expressions that relate the partition function to the following:

- Internal energy
- Helmholtz free energy
- Pressure
- Entropy
- Heat capacity at constant volume

10.1 Definition of the Partition Function

If we consider the expression for the Boltzmann distribution given in equation (9.9), it can be seen that the same denominator appears in the formula regardless of which level's probability is being determined. This term, which is just the sum over all levels of the exponential weighting for that level, is given the name of the total partition function, Q:

$$Q = \sum_{i=1}^{\text{all levels}} \exp\left(-\frac{U_i}{k_B T}\right) \tag{10.2}$$

The partition function is actually the most important quantity in statistical mechanics. It contains the information about how the levels are occupied in a system. We can also attach a physical significance to it:

> *"The partition function represents the number of thermally accessible energy levels at a given temperature."*

We will return to this point again when we begin to actually calculate partition functions for specific cases.

Here we should introduce a piece of notation that is widely used for brevity:

$$\beta = \frac{1}{k_B T} \tag{10.3}$$

Therefore the partition function can also be written as:

$$Q = \sum_{i=1}^{\text{all levels}} \left(-\beta U_i\right) \tag{10.4}$$

The significance of the partition function only becomes apparent when we see that the key thermodynamic quantities can be expressed in terms of Q. In the remainder of this chapter we will demonstrate how this is possible.

10.2 The Internal Energy

From what has gone before, we can write the equation that defines the overall internal energy for a system, introducing Q for the first time:

$$U = \sum_i p_i U_i = \sum_i U_i \frac{\exp(-U_i / k_B T)}{Q} \tag{10.5}$$

From here on it will be implicit that the summations are over all energy levels, unless stated otherwise.

Box 10.1 Differentiation: Revision

The following are a few concepts from differential calculus that will be needed in order to follow the derivations in this chapter:

1. *Differentiation of powers of x:*

$$\frac{\mathrm{d}}{\mathrm{d}x}\left(x^m\right) = m \times x^{(m-1)} \tag{10.6}$$

2. *Differentiation of a natural logarithm:*

$$\frac{\mathrm{d}}{\mathrm{d}x}\left(\ln x\right) = \frac{1}{x} \tag{10.7}$$

3. *Differentiation of a product:*

$$\frac{\mathrm{d}}{\mathrm{d}x}\left(uv\right) = u\frac{\mathrm{d}v}{\mathrm{d}x} + v\frac{\mathrm{d}u}{\mathrm{d}x} \tag{10.8}$$

4. *Chain rule:*

$$\frac{\mathrm{d}}{\mathrm{d}x}\left(f(y)\right) = \frac{\mathrm{d}f}{\mathrm{d}y}\frac{\mathrm{d}y}{\mathrm{d}x} \tag{10.9}$$

We wish to try to rewrite equation (10.5) in terms of Q alone. It is easiest to proceed in a slightly roundabout fashion, by first calculating the first derivative of the partition function with respect to temperature:

$$\frac{\mathrm{d}Q}{\mathrm{d}T} = \sum_i \left(\frac{U_i}{k_\mathrm{B}T^2}\right)\exp\left(-\frac{U_i}{k_\mathrm{B}T}\right) \tag{10.10}$$

Dividing both sides by Q and rearranging, we obtain:

$$k_\mathrm{B}T^2\frac{1}{Q}\frac{\partial Q}{\partial T} = \sum_i U_i \frac{\exp(-U_i / k_\mathrm{B}T)}{Q} \tag{10.11}$$

Comparison with equation (10.5) shows that the right-hand side of equation (10.11) is the result we were seeking for the ensemble average of the internal energy. By using the result of equation (10.7), we can make one final rearrangement:

$$U = k_\mathrm{B}T^2\left(\frac{\partial \ln Q}{\partial T}\right)_V \tag{10.12}$$

Notice that here we have introduced the fact that the partial differential of ln Q with respect to temperature is for constant volume.

There is a small additional consideration for equation (10.12). We are measuring the energy with respect to its value at absolute zero (*i.e.* the ground state energy). If the system contains vibrational degrees of freedom, there will also be a zero-point energy, U_0. Hence we should more correctly write:

$$U - U_0 = k_B T^2 \left(\frac{\partial \ln Q}{\partial T} \right)_V \qquad (10.13)$$

The above equation tells us that the internal energy depends on the rate at which the higher energy levels become occupied with temperature.

10.3 The Helmholtz Free Energy

Earlier we have seen the *Gibbs–Helmholtz* relationship (equation 5.14) between the Gibbs free energy and the enthalpy. If the same derivation is performed under conditions of constant volume rather than constant pressure, an analogous equation connecting the *Helmholtz free energy* and the *internal energy* may be arrived at:

$$\left(\frac{\partial}{\partial T} \left(\frac{A}{T} \right) \right)_V = -\frac{U}{T^2} \qquad (10.14)$$

The statistical mechanical expression for A can now be obtained by simply substituting equation (10.12) into the above expression:

$$\left(\frac{\partial}{\partial T} \left(\frac{A}{T} \right) \right)_V = -\frac{k_B T^2}{T^2} \left(\frac{\partial \ln Q}{\partial T} \right)_V \qquad (10.15)$$

Cancelling the factors of T^2 on the right-hand side, and integrating both sides with respect to temperature at constant volume:

$$A - A_0 = -k_B T \ln Q \qquad (10.16)$$

Once again we have introduced a term, this time A_0, which allows for the value of the free energy at absolute zero and arises as a constant of integration. However, this term is usually zero, except in the case of vibrational energy levels.

10.4 The Entropy

Having derived expressions for the internal energy and Helmholtz free energy in terms of the partition function, it is now straightforward to arrive at one for the entropy by using;

$$A = U - TS \qquad (10.17)$$

which on rearranging gives:

$$S = \frac{U - A}{T} \qquad (10.18)$$

Substituting for U and A from equations (10.13) and (10.16), respectively, yields:

$$S = k_B T\left(\frac{\partial \ln Q}{\partial T}\right)_V + k_B \ln Q \qquad (10.19)$$

Note that the constants U_0 and A_0 are in fact the same, and therefore cancel out, so that the entropy at absolute zero really is zero.

10.5 The Pressure

We can determine a relationship between the partition function and pressure from the equation that connects the pressure to the Helmholtz free energy:

$$p = -\left(\frac{\partial A}{\partial V}\right)_T = k_B T\left(\frac{\partial \ln Q}{\partial V}\right)_T \qquad (10.20)$$

The pressure can be related to the Helmholtz free energy by taking partial derivatives:

$dA = dU - TdS - SdT$

But for a reversible change:

$dU = dq + dw$
$= TdS - pdV$

Combining equations we obtain:

$dA = -SdT - pdV$

Therefore:

$\left(\frac{\partial A}{\partial T}\right)_V = -S; \quad \left(\frac{\partial A}{\partial V}\right)_T = -p$

10.6 The Isochoric Heat Capacity

The final property we shall consider, for the moment, is the *isochoric heat capacity*. This quantity is important since it reflects how easy it is to excite the system to a higher energy state. The heat capacity at constant pressure is more complex to determine as it would involve working in a different ensemble, and therefore we will not consider it here.

Starting from the definition of the heat capacity at constant volume:

$$C_V = \left(\frac{\partial U}{\partial T}\right)_V \qquad (10.21)$$

we can again substitute for U using equation (10.13) to arrive at:

$$C_V = 2k_B T\left(\frac{\partial \ln Q}{\partial T}\right)_V + k_B T^2\left(\frac{\partial^2 \ln Q}{\partial T^2}\right)_V \qquad (10.22)$$

From this we can see that the heat capacity differs from the other properties we have considered so far because it is the only one that involves a second derivative of the partition function. The way in which this quantity differs can be further explored by returning to our original definition of the internal energy as an ensemble average:

$$C_V = \frac{\partial}{\partial T}\left[\frac{\sum_i U_i \exp(-U_i / k_B T)}{\sum_i \exp(-U_i / k_B T)}\right] \qquad (10.23)$$

This differentiation is given as one of the problems below. Here we shall go straight to the result, which is that the heat capacity can be written in the following alternative form:

$$C_V = \frac{\langle U^2 \rangle - \langle U \rangle^2}{k_B T^2} \tag{10.24}$$

where the averages are defined as:

$$\langle U^2 \rangle = \sum_i U_i^2 \frac{\exp(-U_i / k_B T)}{Q} \tag{10.25}$$

$$\langle U \rangle = \sum_i U_i \frac{\exp(-U_i / k_B T)}{Q} \tag{10.26}$$

From equation (10.24), we can see that the heat capacity is related to the difference between the mean squared internal energy and the square of the mean internal energy. The numerator can be further rewritten by introducing the following:

$$\delta U = U - \langle U \rangle \tag{10.27}$$

$$\langle (\delta U)^2 \rangle = \langle U^2 \rangle - \langle U \rangle^2 \tag{10.28}$$

Consequently, the heat capacity is related to the mean square deviation of the internal energy from its average value. In Figure 10.1 a typical distribution of internal energies in the canonical ensemble is illustrated. The heat capacity is an example of a fluctuation property, where the magnitude is a reflection of the width of the distribution of the internal energy about its mean.

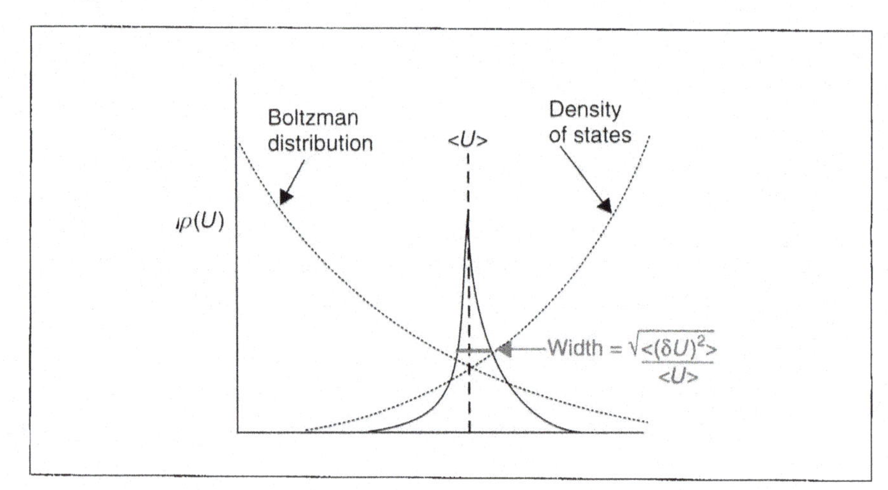

Figure 10.1 The distribution of energies in the canonical ensemble

Summary of Key Points

1. *Definition of the partition function*
   The partition function is defined as:

$$Q = \sum_i \exp\left(-U_i / k_B T\right)$$

2. *Interpretation of the partition function*
   The magnitude of the partition function represents the number of thermally accessible energy levels at a given temperature.

3. *Macroscopic properties and the partition function*
   The measurable value of macroscopic properties, such as U, A, S, p and C_V, can be expressed in terms of the partition function, the temperature and Boltzmann's constant.

Further Reading

R. Bowley and M. Sanchez, *Introductory Statistical Mechanics*, 2nd edn., Oxford University Press, Oxford, 1999, chapter 5.
C .E. Hecht, *Statistical Thermodynamics and Kinetic Theory*, Freeman, New York, 1990, chapter 1.

Problems

1. By differentiating the expression for the heat capacity at constant volume given in equation (10.23), derive the relationship for this property given in equation (10.24).

2. Show that the heat capacity at constant volume can be expressed as:

$$C_V = \frac{\langle (\delta U)^2 \rangle}{k_B T^2}$$

11
An Ideal Gas of Atoms

In our discussion of statistical mechanics so far we have necessarily been quite general. However, we are now in the position where we can consider how to apply these general ideas to specific cases. Arguably, the simplest system we can think of in chemistry is an ideal gas consisting of *single atoms*. The statistical mechanics of this material will be the topic of this chapter.

Aims

In this chapter we will examine how to calculate the properties of an ideal gas using statistical mechanics. By the end of this you should be able to:

- State how the partition function of a collection of ideal gas atoms is related to that of a single atom.
- Calculate the contribution of translation motion to the partition function for an ideal gas.
- Demonstrate that the statistical mechanical results for the energy and pressure are consistent with the previous known classical results.
- Calculate the translational contribution to the entropy of an ideal gas.

11.1 The Ideal Gas

The starting point for this chapter is to consider an ideal gas composed of many individual atoms. In subsequent chapters we shall see how to extend this to the more general situation of an ideal gas consisting of molecules. First we must define what we mean by an ideal gas:

"An ideal gas is one in which the energy of every component atom or molecule is independent of any other."

To express this formally, if the energy of an individual atom i in a state v is $u_v(i)$, then the total internal energy for that state is given by:

$$U_v = u_v(1) + u_v(2) + u_v(3) + \ldots + u_v(N) \tag{11.1}$$

$$U_v = \sum_{i=1}^{N} u_v(i) \tag{11.2}$$

For an ideal gas of atoms the internal energy will consist entirely of *kinetic energy* (if we neglect the electronic energy of the atoms for now). The implication of equation (11.1) is that all atoms move independently from each other in an ideal gas, with no interaction unless they collide.

Although the approximation of a non-interacting ideal gas sounds like a rather crude one, in reality many of the results that follow from this turn out to be surprisingly good in comparison with experiment, at least at low pressures. Statistical mechanics is equally as valid for interacting systems. However, the theory involved is more complex, and beyond the scope of this introductory text.

Now let us consider what happens when we express the partition function in terms of the individual particle energies by substituting equation (11.1) into equation (10.4):

$$Q = \sum_{v} \exp(-\beta U_v) = \sum_{v} \exp\left(-\beta\left(u_v(1) + u_v(2) + \ldots + u_v(N)\right)\right) \tag{11.3}$$

Box 11.1 Mathematical Aside

An important property of exponentials that we now need to recall is the following:

$$\exp(a+b) = \exp(a) \times \exp(b) \tag{11.4}$$

Consequently when taking natural logarithms we also arrive at:

$$\ln(a \times b) = \ln(a) + \ln(b) \tag{11.5}$$

Using the result of equation (11.4), we can manipulate equation (11.3):

$$Q = \left(\sum_{v} \exp\left(-\beta u_v(1)\right)\right)\left(\sum_{v} \exp\left(-\beta u_v(2)\right)\right)\ldots\left(\sum_{v} \exp\left(-\beta u_v(N)\right)\right) \tag{11.6}$$

However, if all the atoms are the same, then so will be the partition function for each atom:

$$Q = \left(\sum_{r} \exp(-\beta u_r) \right)^N = q^N \qquad (11.7)$$

Here we have introduced a new quantity, q, which is the atomic or molecular partition function, depending on which type of particle we are concerned with. As the name suggests, this is just the partition function for an *individual atom or molecule.* To distinguish it from the quantity Q, which is the partition function for the overall system, we recall that this is the total partition function

In fact, the situation is slightly more complicated, since equation (11.7) only applies to particles that are distinguishable. For example, the atoms in a *solid* are associated with particular lattice sites about which they vibrate. Therefore if we label all the atoms we can distinguish them through the coordinates of their position. If we consider the situation in a *liquid* or a *gas*, then the atoms are free to move and swap positions, which makes the atoms indistinguishable from each other.

Consider the system illustrated in Figure 11.1. Here we have three identical atoms except for the arbitrary labels. Although we can draw six different arrangements or complexions of the system, based on the labels that are attached, these are all really identical and indistinguishable. Hence, if we were to include them all, we would have overcounted the number of unique states by 6 (which is derived from 3!). For the above system, we have to divide by six when calculating the total partition function.

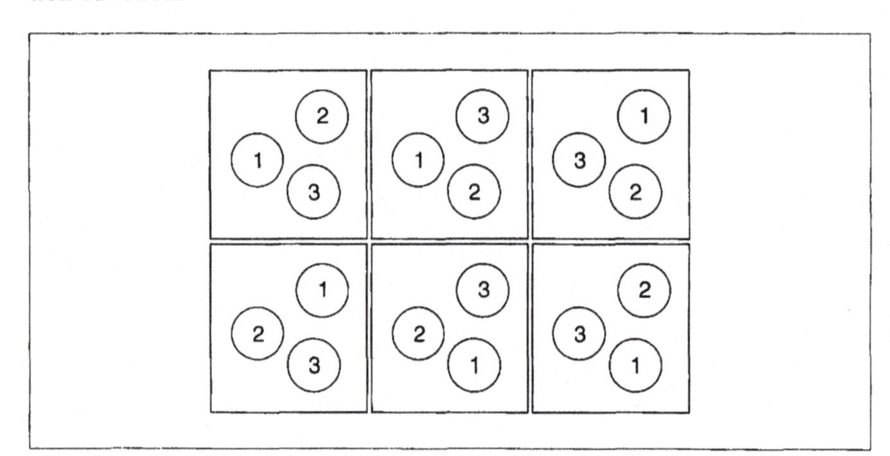

Figure 11.1 The six possible configurations of three gaseous atoms that are indistinguishable

In general, for a system containing N identical atoms or molecules:
Indistinguishable particles (gas/liquid):

$$Q = \frac{q^N}{N!} \qquad (11.8)$$

Distinguishable particles (solid):

$$Q = q^N \tag{11.9}$$

Remember that atoms can be different through both position and type, if we were to consider a mixture.

Worked Problem 11.1

Q A gas is composed of three atoms of type A and four atoms of type B, which have atomic partition functions of q_A and q_B, respectively. What is the total partition function for this system?

A We have to consider the two types of atoms separately, since A and B are distinguishable from each other. Hence we obtain:

$$Q = \left(\frac{q_A^3}{3!}\right)\left(\frac{q_B^4}{4!}\right) = \frac{q_A^3 q_B^4}{144} \tag{11.10}$$

11.2 The Molecular Partition Function

An individual molecule in an ideal gas has energy due to its types of motion – translation (T), rotation (R) and vibration (V) – as well as that due to its electronic state (E). If we make the approximation that the energy levels for these four different processes are independent:

$$u_r = u_r^T + u_r^R + u_r^V + u_r^E \tag{11.11}$$

then we can factorize the molecular partition function into contributions from each form of energy:

$$\begin{aligned} q &= \sum_r \exp\left(-\beta(u_r^T + u_r^R + u_r^V + u_r^E)\right) \\ &= \left(\sum_r \exp\left(-\beta u_r^T\right)\right)\left(\sum_r \exp\left(-\beta u_r^R\right)\right)\left(\sum_r \exp\left(-\beta u_r^V\right)\right)\left(\sum_r \exp\left(-\beta u_r^E\right)\right) \\ &= q^T q^R q^V q^E \end{aligned} \tag{11.12}$$

In this chapter we are considering primarily an ideal monatomic gas, and therefore the challenge is to calculate the translational partition function, q^T, since translation is the only form of motion possible.

11.3 The Translational Partition Function

In order to calculate the translational partition function we will begin by taking a quantum point of view. Every atom has a wavefunction, $\psi(r)$, that describes the probability (ρ) of finding that particle at any given point in space, according to the Born interpretation

$\rho(r) = \psi^*(r)\psi(r)$

Consider an atom of mass m inside a cubic container of side L. We know that the probability of finding the atom must go to zero when it reaches the side, if it is assumed that the atom cannot escape. Hence we need to solve the Schrödinger equation for a particle translating within these boundary conditions.

The **Schrödinger equation** defines the energy, U, of a system that is not evolving with time in terms of the wavefunction, ψ, and the Hamiltonian, H:

$$H\psi = U\psi$$

The Hamiltonian contains terms that describe the potential energy, V, of the electrostatic interaction of the electrons and nuclei, with themselves and each other, plus the kinetic energy of these particles:

$$\left(\sum_{i=1}^{\text{all particles}} -\frac{h^2}{8\pi^2 m_i}\left(\frac{\partial^2}{\partial x^2}+\frac{\partial^2}{\partial y^2}+\frac{\partial^2}{\partial z^2}\right)+V\right)\psi = U\psi$$

The solution is that of the well-known particle in a box, which yields allowed energy levels, u_v, of:

$$u_v = \frac{(n_x^2 + n_y^2 + n_z^2)h^2}{8mL^2} \tag{11.13}$$

where n_x, n_y, n_z are the quantum numbers in each Cartesian direction and take values between 1 and ∞.

Box 11.2 Particle in a Box

In one dimension, a particle in a potential well of length L with infinite sides can be described by a sine wave (see Figure 11.2):

$$\psi = A\sin\left(\frac{\pi n_x x}{L}\right) \tag{11.14}$$

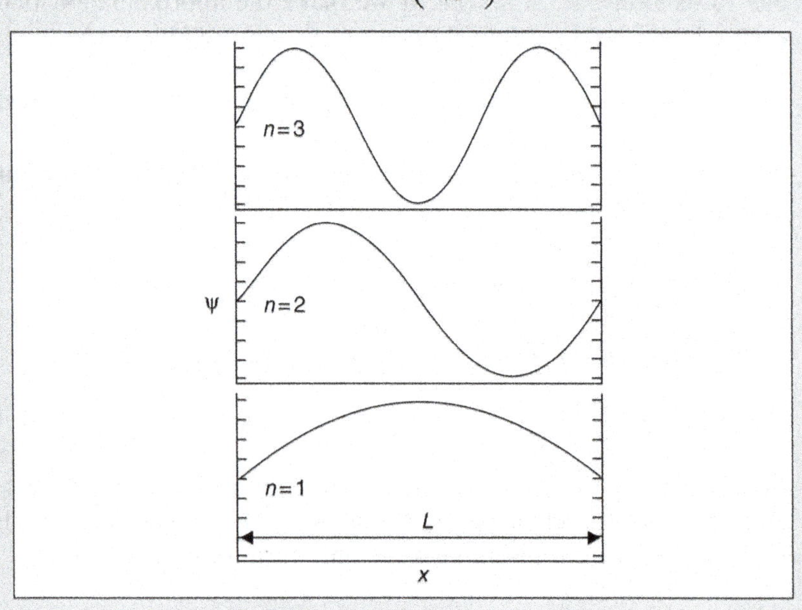

Figure 11.2 The three lowest energy wavefunctions for a particle in a box

As the potential is zero everywhere within the box, we have the simplified Schrödinger equation:

$$\left(-\frac{h^2}{8\pi^2 m}\frac{\partial^2}{\partial x^2}\right)\psi = u\psi \tag{11.15}$$

Solving for u, we obtain:

$$u = \frac{n_x^2 h^2}{8mL^2} \tag{11.16}$$

As n_x increases, so does the number of *nodes* where the wavefunction changes sign in order to maintain *orthogonality* with all other wavefunctions. This increasing *curvature* as the wavefunction tries to accommodate more nodes in a fixed length leads to a higher *kinetic energy* for translation.

Using the particle in a box result, the translational partition function is:

Two wavefunctions, ψ_a and ψ_b, are said to be *orthogonal* if the integral of their product over all space is exactly zero:

$$\oint \psi_a{}^* \psi_b \, d\tau = \int_{-\infty}^{\infty}\int_{-\infty}^{\infty}\int_{-\infty}^{\infty} \psi_a{}^* \psi_b \, dx\, dy\, dz = 0$$

$$q^{\mathrm{T}} = \sum_{n_x}\sum_{n_y}\sum_{n_z} \exp\left(-\frac{(n_x^2 + n_y^2 + n_z^2)h^2}{8mL^2 k_{\mathrm{B}} T}\right) \tag{11.17}$$

where the triple summation is over all combinations of n_x, n_y and n_z. Note that we have summed over all the states, neglecting the fact that many will be *degenerate* due to permutations of the three quantum numbers. Again, the result of equation (11.4) may be used to factorize the expression in equation (11.17):

$$q^{\mathrm{T}} = \left\{\sum_{n_x} \exp\left(-\frac{n_x^2 h^2}{8mL^2 k_{\mathrm{B}} T}\right)\right\}\left\{\sum_{n_y} \exp\left(-\frac{n_y^2 h^2}{8mL^2 k_{\mathrm{B}} T}\right)\right\}\left\{\sum_{n_z} \exp\left(-\frac{n_z^2 h^2}{8mL^2 k_{\mathrm{B}} T}\right)\right\} \tag{11.18}$$

It is clear that for a cubic box the three terms in the product are identical, and therefore:

$$q^{\mathrm{T}} = \left[\sum_{n_x} \exp\left(-\frac{n_x^2 h^2}{8mL^2 k_{\mathrm{B}} T}\right)\right]^3 \tag{11.19}$$

When two, or more, energy levels have the same energy, then they are said to be **degenerate**. For a particle in a box, we can describe the state of the system by specifying the values of n_x, n_y and n_z, in the abbreviated form (n_x,n_y,n_z). From the form of equation (11.17), the energy levels (2,1,1), (1,2,1) and (1,1,2) will be degenerate with each other, for example.

Worked Problem 11.2

Q What is the energy required to excite a hydrogen atom (mass = 1.674×10^{-24} g) in a cubic box of volume 1 m^3 from the ground state ($n_x = 1$, $n_y = 1$, $n_z = 1$) to the first excited state ($n_x = 2$, $n_y = 1$, $n_z = 1$)?

A Because the excitation only involves the x direction we can calculate the energy change by treating the problem as one-dimensional. Hence we can use equation (11.16), with $L = 1$ m and taking care to employ SI units:

$$\Delta u = \frac{(2^2 - 1^2)(6.63 \times 10^{-34}\text{ J s})^2}{8(1.674 \times 10^{-27}\text{ kg})(1\text{ m})^2} = 9.85 \times 10^{-41}\text{ J (to 3 s.f.)}$$

Comment: The spacing between translational energy levels is generally found to be negligible.

In order to further develop the translational partition function we need to make two more steps:

Approximations:

- The translational energy levels are so close together that the integer quantum number, n_x, can be converted to a continuous variable. In other words, the summation in equation (11.19) can be replaced by an integral.
- The integral of n_x from 1 to ∞ can be replaced by the integral from 0 to ∞ with negligible effect. This allows a standard integral to be used.

Using these approximations, we arrive at the following:

$$q^{\mathrm{T}} = \left[\int_0^{\infty} \exp\left(-\frac{n_x^2 h^2}{8mL^2 k_{\mathrm{B}} T}\right) \mathrm{d}n_x\right]^3 \quad (11.20)$$

Many exponential functions of variables between the limits of 0 and ∞ take the form of standard integrals, which can be found in mathematical tabulations:

$$I = \int_0^{\infty} \exp(-ax^2)\mathrm{d}x = \frac{1}{2}\sqrt{\frac{\pi}{a}}$$

Solving equation (11.20) with a standard integral leads to the simplified expression for the translational partition function:

$$q^{\mathrm{T}} = \left(\frac{2\pi m k_{\mathrm{B}} T}{h^2}\right)^{3/2} V \quad (11.21)$$

where V is the volume, equal to L^3. The thermal de Broglie wavelength, λ, of a particle of mass m and a temperature T, is given by:

$$\lambda = \left(\frac{h^2}{2\pi m k_B T} \right)^{1/2} \tag{11.22}$$

The expression for the translational partition function can therefore be rewritten in terms of the de Broglie wavelength by substituting equation (11.22) into equation (11.21):

$$q^T = \lambda^{-3} V \tag{11.23}$$

In order to calculate the translational partition function we therefore need the following pieces of information:

- m, the mass of a single atom or molecule (kg)
- T, the temperature (K)
- V, the volume that the gas occupies (m^3)

Worked Problem 11.3

Q Evaluate q^T for an oxygen molecule (molecular weight 31.99 g mol^{-1}) at 273.15 K when occupying a volume of 22.414 dm^3.

A The first task is to convert the mass per mole of molecules into the mass per molecule in kg:

$$m = \frac{31.99}{N_A} \text{ g} = 5.312 \times 10^{-23} \text{ g} = 5.312 \times 10^{-26} \text{ kg}$$

Now equation (11.22) can be employed, provided that the units of the volume are changed to m^3:

$$q^T = \left(\frac{2\pi(5.312 \times 10^{-26})(1.38066 \times 10^{-23})(273.15)}{(6.63 \times 10^{-34})^2} \right)^{3/2} (0.022414)$$
$$= 3.26 \times 10^{30} \quad \text{(to 3 s.f.)}$$

11.4 The Internal Energy of a Monatomic Ideal Gas

We are now in a position to write down the total partition function for a monatomic gas containing N atoms in terms of the atomic translational partition function:

$$Q = \frac{q^N}{N!} = \frac{\lambda^{-3N} V^N}{N!} \tag{11.24}$$

Substituting this into the relationship between the total partition function and the *internal energy* from equation (10.12) yields the following expression:

$$U = k_{\mathrm{B}}T^2 \left(\frac{\partial \ln(\lambda^{-3N} V^N / N!)}{\partial T} \right)_V \tag{11.25}$$

To simplify matters we can use the properties of logarithms, as shown in equation (11.5), to separate any terms that depend on T. The derivative with respect to T of any remaining terms will then be zero and can be neglected. In equation (11.22), only λ depends on T, and so we can reduce this to:

$$U = k_{\mathrm{B}}T^2 \left(\frac{\partial \ln T^{3N/2}}{\partial T} \right)_V \tag{11.26}$$

$$U = k_{\mathrm{B}}T^2 \left(\frac{3N}{2} \frac{\partial \ln T}{\partial T} \right)_V \tag{11.27}$$

$$U = \frac{3N}{2} k_{\mathrm{B}}T^2 \left(\frac{1}{T} \right) \tag{11.28}$$

$$U = \frac{3}{2} N k_{\mathrm{B}} T \tag{11.29}$$

To specify exactly where an atom is in space requires three independent Cartesian coordinates, x, y and z. Each of these is a **degree of freedom**. When atoms are part of a molecule, the degrees of freedom take the form of **translation**, **rotation** and **vibration**. However, the total number of degrees of freedom for a molecule containing N atoms will always be $3N$. The distribution of degrees of freedom between types of motion for $3N$ atoms can be summarized as follows:

	Translation	*Rotation*	*Vibration*
Atoms	$3N$	0	0
Linear molecule	3	2	$3N - 5$
Non-linear molecule	3	3	$3N - 6$

While this is the final result for N atoms, we can write a simpler form for the molar translational internal energy by using the relationship between Boltzmann's constant and the Universal gas constant, $R = N_{\mathrm{A}}k_{\mathrm{B}}$:

$$U_{\mathrm{m}} = \frac{3}{2} RT \tag{11.30}$$

The above results for the internal energy are exactly the values that arise from classical theory according to the equipartition theorem:

The equipartition theorem states that every degree of freedom should have an energy of $\frac{1}{2} k_B T$.

Since there are three translational degrees of freedom per atom, it can be seen that this is consistent with equation (11.29). The reason that we have arrived at a classical result, despite beginning from quantized energy levels, is because we replaced a *summation* by an *integral* in deriving the expression for the translational partition function. An integral requires that a variable be continuous, which in this case implies that there are no discrete energy levels. This means that energy can be added, or removed, from a system in any arbitrary amount, which is consistent with the classical view of energy.

11.5 The Heat Capacity of a Monatomic Ideal Gas

From the result for the internal energy we can derive an expression for the molar heat capacity at constant volume of an ideal gas, which is again equivalent to the classical result:

$$C_V = \left(\frac{\partial U_\mathrm{m}}{\partial T}\right)_V = \frac{3}{2}R \tag{11.31}$$

11.6 The Pressure of a Monatomic Ideal Gas

We recall that the pressure is given by the expression:

$$p = k_\mathrm{B}T\left(\frac{\partial \ln Q}{\partial V}\right)_T \tag{11.32}$$

Again we can make use of the properties of the logarithm, as given in equation (11.5), so that we need only consider the terms in the translational partition function that depend on volume, since all others will vanish when differentiated:

$$p = k_\mathrm{B}T\left(\frac{\partial \ln V^N}{\partial V}\right)_T = k_\mathrm{B}T\left(\frac{1}{V^N}\right)(NV^{N-1}) = \frac{Nk_\mathrm{B}T}{V} \tag{11.33}$$

This final result can be rewritten in the more familiar form:

$$pV = nRT \tag{11.34}$$

which is the ideal gas equation. Hence statistical mechanics offers a means of deriving the *classical* relationship between the pressure, volume, number of atoms and temperature of an ideal gas.

11.7 The Entropy of a Monatomic Ideal Gas

So far, all of the derivations of the properties of an ideal gas have led to well-known classical results. However, no equivalent formula exists for the *entropy* of an ideal gas. Consequently, the prediction of the entropy represents the first real test of statistical mechanics. Again we begin by recalling the definition of the entropy in terms of the partition function:

$$S = k_\mathrm{B}T\left(\frac{\partial \ln Q}{\partial T}\right)_V + k_\mathrm{B}\ln Q \tag{11.35}$$

Substituting for the total translational partition function and neglecting variables other than temperature in the first term of the entropy:

$$S = k_B T \left(\frac{\partial \ln T^{3N/2}}{\partial T} \right)_V + k_B \ln \left[\left(\frac{2\pi m k_B T}{h^2} \right)^{3N/2} \frac{V^N}{N!} \right] \qquad (11.36)$$

Box 11.3 Mathematical Aside

In order to be able to handle the logarithm of the factorial of a large number, it is useful to use Stirling's approximation, which is valid for large N:

$$\ln N! \approx N \ln N - N \qquad (11.37)$$

When the number of atoms is large, we can modify equation (11.36) by substituting in (11.37), while also evaluating the derivative with respect to temperature:

$$S = \frac{5Nk_B}{2} + Nk_B \ln \left[\frac{(2\pi m k_B T)^{3/2} V}{Nh^3} \right] \qquad (11.38)$$

For an ideal gas it is more usual to express its state in terms of its pressure, rather than the volume it occupies. Thus we can rewrite equation (11.38) by eliminating the volume through the use of the ideal gas equation. To further simplify matters we can also express it as the *molar translational entropy of an ideal gas*:

$$S_m = \frac{5R}{2} + R \ln \left[\frac{(2\pi m k_B T)^{3/2} RT}{N_A h^3 p} \right] \qquad (11.39)$$

This is known as the Sackur–Tetrode equation

Worked Problem 11.4

Q Calculate the molar entropy of argon gas ($M = 39.948$ g mol^{-1}) at 298.15 K and 1 bar of pressure (1 bar = 10^5 Pa; 1 Pa = 1 N m^{-2}).

A The Sackur–Tetrode equation can be used for this problem, taking care as always to convert all quantities to SI units:

$$S_m = \frac{5R}{2} + R\ln\left[\frac{\left(2\pi\left(\frac{39.948\times10^{-3}}{N_A}\right)k_B(298.15)\right)^{3/2}}{N_A h^3}\frac{R(298.15)}{10^5}\right]$$

$$= 20.785 + 8.314\ln\left[\frac{(1.716\times10^{-45})^{3/2}}{1.755\times10^{-76}}2.479\times10^{-2}\right]\text{J K}^{-1}\ \text{mol}^{-1}$$

$$= 20.785 + 8.314\ln\left[1.0038\times10^{7}\right] = 154.8\ \text{J K}^{-1}\ \text{mol}^{-1}$$

Comment: The value measured by calorimetry is 154.6 J $K^{-1}mol^{-1}$, which is in good agreement with our calculated estimate of the entropy.

In Table 11.1, a selection of entropy values for gases is given, as predicted from statistical mechanics and also measured experimentally. The fact that the values agree to within the estimated experimental error was an early success for statistical mechanics and led to credibility of the hypotheses previously introduced.

Table 11.1 The entropy of gases at one atmosphere pressure (~10^5 Pa)

Gas	*T*/K	$S_{predicted}$/J K^{-1} mol^{-1}	$S_{measured}$/J K^{-1} mol^{-1}
Cd	298.1	154.8	152.3
Hg	298.1	167.8	167.4
N_2	298.1	161.1	160.7
O_2	298.1	205.0	205.4
HCl	298.1	186.6	186.2
HBr	298.1	198.7	199.2
NH_3	239.7	184.5	184.5
CO_2	194.7	198.7	199.2

Summary of Key Points

1. *The total and molecular partition functions*
   The total partition function for a collection of *N* molecules or atoms can be expressed as the product of the partition functions for the individual species. If the atoms or molecules are

indistinguishable owing to translational motion, the total partition function must be divided by a factor of $N!$

2. *Contributions to the partition function of an ideal gas*
   The molecular partition function for an ideal gas, where the energy is just the sum of contributions, can be expressed as the product of the partition function for four common types of energy: translational, rotational, vibrational and electronic.

3. *The translational partition function*
   The partition function for translational motion can be derived from the particle in a box model, and is found to be a function of the mass of the molecule or atom, the temperature, and the volume occupied by the gas.

4. *The internal energy and pressure for translation*
   Because the spacing between translational energy levels is normally so small as to be negligible, statistical mechanics leads to the classical results for the internal energy ($\frac{1}{2}k_BT$ per degree of freedom) and the pressure ($pV = nRT$).

5. *The molar entropy of an ideal gas*
   Calculated molar entropies for gases from the statistical mechanical formula agree well with experimental measurements at ambient pressure. This offers the first verification of the hypotheses previously introduced.

Further Reading

P. W. Atkins, *Physical Chemistry*, 6th edn., Oxford University Press, Oxford, 1998, chapter 19.

R. Bowley and M. Sanchez, *Introductory Statistical Mechanics*, 2nd edn., Oxford University Press, Oxford, 1999, chapter 5.

D. Chandler, *Introduction to Modern Statistical Mechanics*, Oxford University Press, Oxford, 1987, chapter 4.

D. A. McQuarrie and J. D. Simon, *Molecular Thermodynamics*, University Science Books, Sausalito, California, 1999.

Problems

1. By what factor would the total partition function for a system of four atoms in the solid state change if melting occurred?

2. Two gaseous systems consist of mixtures of atoms of type A and B in containers of equal volume, where the atomic partition functions are q_A and q_B, respectively. The first system consists of four atoms of A and two atoms of B, while the second contains three atoms of both types. Write down expressions for the total partition functions of (a) both separate systems, (b) the system that would result from mixing both systems in the container of one of them, (c) the combined system that arises by linking both containers together.

3. Calculate (a) the de Broglie wavelength and (b) the translational partition function for an atom of Br (mass = 79.90 g mol^{-1}) at 500 K when occupying a volume of 35 dm^3.

4. Derive an expression for the translational contribution to the Helmholtz free energy of an ideal gas.

5. Calculate the molar entropy of Zn (molar mass = 65.38 g mol^{-1}) as an ideal gas at a temperature of 800 K and a pressure of 1 GPa.

12
An Ideal Gas of Diatomic Molecules

In order to keep everything as simple as possible, we have so far only considered in detail the statistical mechanics of *atoms*. In this chapter we will take the next step of examining the case of an *ideal gas* composed of *diatomic molecules*. As a consequence, it is necessary to introduce the calculation of the *rotational* and *vibrational* partition functions. Finally, we shall briefly discuss how to evaluate also the *electronic* contribution to the partition function.

Aims

In this chapter we will demonstrate how to evaluate, using statistical mechanics, the partition function and resulting properties of an ideal gas composed of diatomic molecules. By the end of this you should be able to:

- Derive an expression for the rotational partition function of a diatomic molecule
- Explain what the symmetry number is and how it arises
- Evaluate the rotational contribution to simple thermodynamic properties for diatomic molecules
- Derive an expression for the vibrational partition function of a harmonic oscillator
- Calculate the energy and heat capacity of vibration according to statistical mechanics
- Plot and explain the variation of the energy and heat capacity with temperature for a harmonic oscillator
- Evaluate the electronic partition function, given the energies and degeneracies of the electronic energy levels

12.1 The Rotational Partition Function

As soon as we move from considering an ideal gas that consists only of atoms to one that is composed of diatomic molecules, we gain two additional forms of motion: *rotation* and *vibration*. For a diatomic molecule there will be a total of $3N = 6$ degrees of freedom, of which 3 are translations, 2 are rotations and 1 is a vibration (recall the margin note on page 92 in Chapter 11).

First of all, let us consider the case of a rotating molecule composed of two atoms with masses m_1 and m_2 as shown in Figure 12.1. The rotational properties of such a molecule are determined by its moment of inertia, I, which can be expressed in terms of the masses and their distances from the centre of mass

$$I = \sum_{i=1}^{N} m_i r_i^2 \tag{12.1}$$

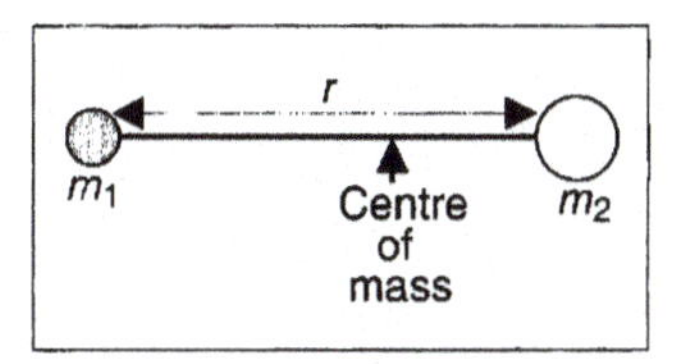

Figure 12.1 A heteronuclear diatomic molecule

However, for the case of a *diatomic molecule only*, there is a simplified formula:

$$I = \mu r^2 \tag{12.2}$$

where r is now the bond length in the molecule and μ is the reduced mass of the system, which is defined as:

$$\mu = \frac{m_1 m_2}{m_1 + m_2} \tag{12.3}$$

Hence in the simple case of a homonuclear diatomic molecule, where both atoms are the same (*e.g.* N_2), then the reduced mass is just half the mass of a single atom.

If the masses of the atoms are in g mol^{-1}, then the reduced mass in kg is:

$$\mu = \frac{m_1 m_2}{m_1 + m_2} \frac{10^{-3}}{N_A} \text{ kg}$$

Solving the Schrödinger equation for a freely rotating molecule yields the following expression for the allowed rotational energy levels:

$$u_J^{\mathrm{R}} = BJ(J+1) \tag{12.4}$$

where B is the rotational constant of the molecule, which is a function of the moment of inertia and fundamental constants:

$$B = \frac{h^2}{8\pi^2 I} \tag{12.5}$$

Strictly speaking, the rotational constant is not actually a constant, since the bond length is also a function of the state of the molecule with regard to other forms of motion. However, we shall ignore such complications here, and assume that we are dealing with a so-called rigid rotor where the bond length is fixed.

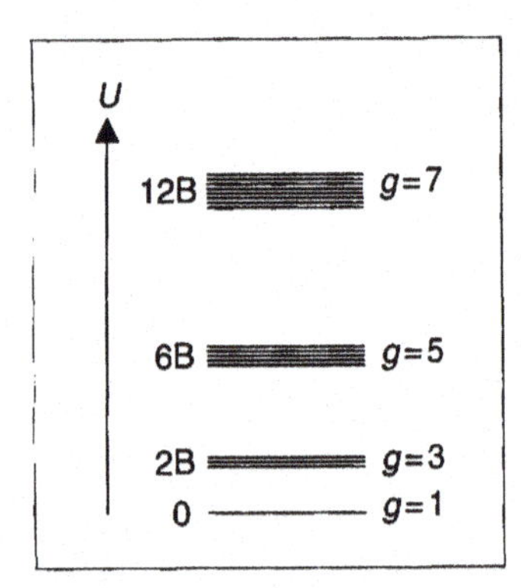

Figure 12.2 The rotational energy levels for a diatomic molecule

The quantity J is the rotational quantum number, and takes integer values from 0 upwards. The sequence of rotational energy levels is illustrated in Figure 12.2. Each of the rotational energy levels also has a degeneracy, g, of $(2J + 1)$.

When degeneracies are present, we can re-write the expression for the partition function, taking only the rotational contribution to equation (11.12), so that we only sum over unique energies and just multiply by the number of degenerate levels:

$$q = \sum_{i=1}^{\text{nondegenerate states}} g_i \exp\left(-\frac{u_i}{k_B T}\right) \tag{12.6}$$

Hence for the case of rotation we can write down the expression that defines the rotational partition function by combining equations (12.4) and (12.5) to obtain the energy, u_i, and substituting into equation (12.6):

$$q^R = \sum_{J=0}^{\infty} (2J+1) \exp\left(-\frac{J(J+1)h^2}{8\pi^2 I k_B T}\right) \tag{12.7}$$

As was the case for translational motion, the spacing of the rotational energy levels is generally quite small in comparison with $k_B T$, which means that the above summation can again be replaced by an integral to simplify the result. However, care must be taken as this will only be *valid* for molecules with *large moments of inertia, i.e.* those which contain heavy atoms or have long bond lengths. In general we should consider the magnitude of the energy level separation, ΔU, relative to thermal energies.

Equation (12.7) can actually be rewritten by introducing a quantity referred to as the rotational temperature, θ_r, that collects together constants which have the combined units of Kelvin:

$$q^R = \sum_{J=0}^{\infty} (2J+1) \exp\left(-\frac{J(J+1)\theta_r}{T}\right) \tag{12.8}$$

$$\theta_r = \frac{h^2}{8\pi^2 I k_B} \tag{12.9}$$

The rotational temperature turns out to be particularly useful in trying to decide whether a molecule will show classical or quantum behaviour, depending on the temperature:

- *Classical behaviour*: $\Delta U << k_B T$ $T >> \theta_r$ => integral
- *Quantized behaviour:* $\Delta U >> k_B T$ $T << \theta_r$ => summation

Assuming that we are in the classical regime, we can now proceed to replace the summation by an integral. To do this requires a substitution of variables with $x = J(J + 1)$ and therefore $\mathrm{d}x = (2J + 1)\mathrm{d}J$:

$$q^{\mathrm{R}} = \int_0^\infty (2J+1)\exp\left(-\frac{J(J+1)\theta_\mathrm{r}}{T}\right)\mathrm{d}J \tag{12.10}$$

$$q^{\mathrm{R}} = \int_0^\infty \exp\left(-\frac{\theta_\mathrm{r}}{T}x\right)\mathrm{d}x \tag{12.11}$$

$$q^{\mathrm{R}} = \left[-\frac{T}{\theta_\mathrm{r}}\exp\left(-\frac{\theta_\mathrm{r}}{T}x\right)\right]_0^\infty \tag{12.12}$$

Substituting the limits of infinity and zero into the above expression leads to the final result for the rotational partition function:

$$q^{\mathrm{R}} = \frac{T}{\theta_\mathrm{r}} = \frac{8\pi^2 I k_\mathrm{B} T}{h^2} \tag{12.13}$$

Worked Problem 12.1

Q Calculate the rotational partition function for the molecule HD at 298.15 K, given that the moment of inertia for HD is 6.29×10^{-48} kg m^2.

A There are two possible ways of calculating the rotational partition function: either by using the explicit summation or by using the integrated result. As an illustration we shall take both approaches here.

First let us calculate the rotational temperature:

$$\theta_\mathrm{r} = \frac{\left(6.626\times10^{-34}\ \mathrm{J\ s}\right)^2}{8\pi^2\left(6.29\times10^{-48}\ \mathrm{kg\ m^2}\right)\times\left(1.38066\times10^{-23}\ \mathrm{J\ K^{-1}}\right)}$$

which yields a value of 64.03 K (to 2 d.p.). Based on this we can say that at 298.15 K the classical approach should be valid.

Classical result:

$$q^{\mathrm{R}}_{\mathrm{classical}} = \frac{T}{\theta_\mathrm{r}} = \frac{298.15}{64.03} = 4.656$$

Quantum result:

$$q^{\mathrm{R}}_{\mathrm{quantum}} = \sum_{J=0}^{\infty}(2J+1)\exp\left(-\frac{J(J+1)\times64.03}{298.15}\right)$$

$$= 1.000 + 1.952 + 1.379 + 0.532 + 0.123 + 0.017 + 0.002 + ...$$

We can see that the contribution due to each rotational level goes through a maximum at $J = 1$ and then decays away rapidly, becoming negligible by $J = 6$. This is due to the competition between the degeneracy, which increases linearly, and the exponential decay, which dominates at large J. The converged value of the summation is found to be 5.005 (to 3 d.p.).

Comment: The difference between the two results above, despite being above the characteristic rotational temperature, is a reflection of the loss of accuracy due to replacing a summation by an integral when quantization is still significant at this temperature.

12.2 Rotation and Symmetry

Having arrived at the result for the rotational partition function for a diatomic molecule, we now have to consider a complication that arises for rotational motion. There is a subtle problem in that the quantum properties of nuclei cause the occupation of only certain rotational states to be permitted. In the case of diatomic molecules, this complication *only applies to the homonuclear case.*

Nuclei, just like electrons and other particles, can possess spin. We can divide all particles into two categories:

- *Fermions*: particles with half-integral spin (1/2, 3/2, ...)
- *Bosons*: particles with integral spin (0, 1, 2, ...)

Examples of fermions are electrons, protons and isotopes of certain atoms including ^{13}C, while deuterons, photons and atoms such as ^{16}O are bosons.

Just as when considering the electronic structure of atoms, we must allow for the Pauli exclusion principle. In a more general form than that which is usually encountered when discussing the electronic structure of atoms, this can be stated as follows:

> *"When two identical particles are interchanged, the total wavefunction must change sign for fermions and remain unchanged for bosons."*

A **wavefunction**, Ψ, must obey the following conditions when two particles are interchanged:

Fermions: $\Psi \rightarrow -\Psi$
Bosons: $\Psi \rightarrow \Psi$

This selection rule controls which rotational transitions are allowed, depending on the nuclei within the molecule. Before we can apply these rules we need to know what the rotational wavefunctions look like. These correspond to the rotational energy levels obtained by solving Schrödinger's equation (12.4), and have the appearance shown in Figure

12.3 in the case of a homonuclear diatomic molecule for the three lowest-lying levels.

As can be seen, the shape is just the same as the spherical harmonics that correspond to the angular momentum quantum number, l, for the electronic wavefunctions of the hydrogen atom. Hence $J = 0$ resembles an s orbital, $J = 1$ a p orbital and so on.

Now let us consider the effect of rotating a homonuclear diatomic molecule by 180°. For *even J levels*, the rotational wavefunction is *symmetric* and therefore the wavefunction does not change sign. Therefore this would be an allowed rotational state for a molecule where the nuclei were *bosons*, since it complies with the Pauli exclusion principle. By consideration of the rotational symmetry, we can arrive at the general rules:

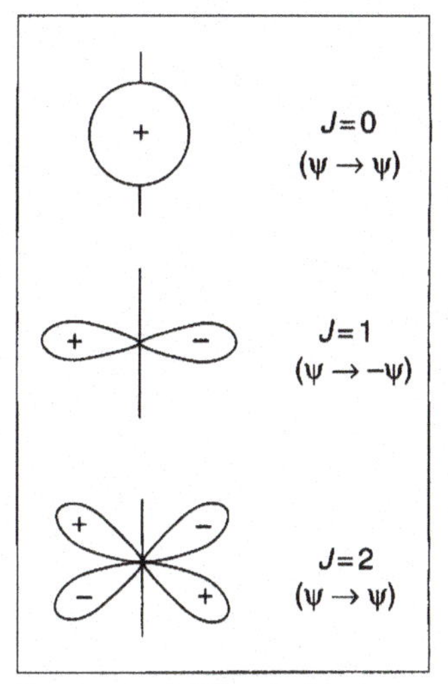

Figure 12.3 The rotational J=0-2 wavefunctions for a homonuclear diatomic molecule

- *Fermions can only exist in rotational levels with an even J value.*
- *Bosons can only exist in rotational levels with an odd J value.*

The main consequence of these selection rules is to influence the levels that we have to include when performing our summation to calculate the rotational partition function. For example, for the molecule $^{16}O_2$, where the nuclei are bosons, the partition function now becomes:

$$q^{\mathrm{R}}(^{16}\mathrm{O}_2) = \sum_{J\,\mathrm{even}}^{\infty} (2J+1)\exp\left(-\frac{J(J+1)h^2}{8\pi^2 I k_{\mathrm{B}} T}\right) \quad (12.14)$$

If we are in the regime where the temperature is large enough relative to the rotational temperature that we can use the integral approximation, then effectively we ignore half the area under the curve and thus just divide the integral by a factor of 2:

$$q^{\mathrm{R}}(^{16}\mathrm{O}_2) = \frac{8\pi^2 I k_{\mathrm{B}} T}{2h^2} \quad (12.15)$$

Indeed, the above formula would apply regardless of whether the nuclei were fermions or bosons, since both must be divided by a factor of two.

In general, whenever a molecule has rotational symmetry, the number of allowed levels is reduced owing to the nuclear spin. Hence we can introduce a so-called symmetry number, σ, which is the factor by which the rotational partition function must be divided to correct for the reduced number of accessible rotational states:

$$q^{\mathrm{R}} = \frac{8\pi^2 I k_{\mathrm{B}} T}{\sigma h^2} \quad (12.16)$$

For diatomic molecules we have have the following rules:

- *Heteronuclear diatomics*: $\sigma = 1$
- *Homonuclear diatomics*: $\sigma = 2$

For more complex molecules than diatomics we require a knowledge of group theory in order to be able to define the symmetry number; this is beyond the scope of this text. However, for completeness we shall state the general definition of the symmetry number:

> *"The symmetry number is the order of the rotation sub-group of the molecule. This is equal to the number of rotational symmetry elements, plus the identity."*

In Table 12.1 we give examples of the symmetry numbers for a few common molecules, showing how this was arrived at from the symmetry elements. However, for a full explanation of the symmetry it is necessary to consult the Further Reading.

Table 12.1 The symmetry numbers of some typical molecules

Molecule	*Rotational sub-group*	σ
HCl	E	1
O_2	E, C_2	2
H_2O	E, C_2	2
NH_3	E, $2C_3$	3
CH_4	E, $8C_3$, $3C_2$	12
C_6H_6	E, $2C_6$, $2C_3$, C_2	6

12.3 The Properties of a Rigid Rotor

For a system of N independent diatomic molecules, when the temperature is sufficiently large, the rotational partition function is:

$$Q = \left(q^{\mathrm{R}}\right)^N = \left(\frac{T}{\sigma\theta_{\mathrm{r}}}\right)^N \tag{12.17}$$

Recall that we do not require a factor of $N!$ because the arguments about distinguishability only apply to translational degrees of freedom.

Using the expression derived earlier (equation 10.12), we can determine the internal energy due to rotation:

$$U^{\mathrm{R}} = k_{\mathrm{B}}T^2 \frac{\partial}{\partial T}\ln\left[\left(\frac{T}{\sigma\theta_{\mathrm{r}}}\right)^N\right] = Nk_{\mathrm{B}}T^2 \frac{\partial}{\partial T}\ln\left(\frac{T}{\sigma\theta_{\mathrm{r}}}\right) \tag{12.18}$$

$$U^{R} = Nk_BT^2 \frac{\partial}{\partial T} \ln T = Nk_BT \quad (12.19)$$

Note that because the symmetry number does not depend on temperature it disappears during the above differentiation. Hence, the result for the internal energy is not influenced by the symmetry of the system and the fact that only some of the levels are accessible to molecules, a result that at first may be surprising.

As with the translational energy, this accords with the classical result from the *equipartition theorem* which states that each degree of freedom has associated with it $\frac{1}{2}k_BT$ worth of energy. For a linear molecule there are just *two* degrees of rotational freedom, rather than the *three* found in the general case. Again this agreement with classical theory arises because we have started from the partition function result obtained by integration (*i.e.* we have explicitly neglected quantization).

From the above result for the internal energy, we can arrive at the expression for the molar heat capacity at constant volume for rotation of a diatomic molecule:

$$C_V^{R} = \frac{\partial}{\partial T}(RT) = R \quad (12.20)$$

Worked Problem 12.2

Q Derive an expression for the rotational contribution to the molar Helmholtz free energy of a diatomic molecule.

A The rotational partition function for a mole of diatomic molecules can be written as:

$$Q = \left(\frac{T}{\sigma\theta_r}\right)^{N_A}$$

Using the definition given earlier for the Helmholtz free energy in equation (10.16):

$$A = A_0 - k_BT \ln Q$$

we can substitute for Q and also take note that the Helmholtz free energy of rotation at absolute zero, A_0, is zero, leading to:

$$A = -N_Ak_BT \ln\left(\frac{T}{\sigma\theta_r}\right) = -RT \ln\left(\frac{T}{\sigma\theta_r}\right)$$

12.4 The Harmonic Oscillator

A diatomic molecule possesses just a single vibration frequency nor-mal mode for the stretching of the bond. This characteristic frequency, ω, can be related to the force constant of the bond, k, and the reduced mass, μ, according to the following relationship:

$$\omega = \frac{1}{2\pi}\sqrt{\frac{k}{\mu}} \tag{12.21}$$

For simplicity we shall only consider the case of the harmonic oscillator, since this leads to more compact expressions for the partition function and thermodynamic properties. This will be a good approximation for most diatomic molecules at low temperatures, since the spacing between levels is usually quite large and therefore only a few low-lying vibrational states will be occupied. However, if we were to consider higher temperatures or more complex molecules with low-frequency motions, then inclusion of anharmonicity would become important.

The solution of the Schrödinger equation for a harmonic oscillator gives rise to a set of equally spaced, singly degenerate levels, as shown in Figure 12.4. The allowed energies are given by the expression:

$$u_n^v = \left(n + \frac{1}{2}\right)h\omega \qquad n = 0, 1, 2... \tag{12.22}$$

where n is the vibrational quantum number. An important feature to note is that even in the $n = 0$ vibrational ground state the molecule possesses a vibrational energy of $\frac{1}{2}h\omega$, which is known as the zero point energy (see Figure 12.4). The reason that this exists is linked to Heisenberg's uncertainity principle, one of the fundamental principles on which quantum mechanics is founded. In one guise this states that it is not possible to know both the exact position and momentum of a particle

The **harmonic oscillator** assumes that the potential energy, V, about the equilibrium bond length, r_0, is given by:

$$V(r) = \frac{1}{2}k(r - r_0)^2$$

Here the increase in potential energy is the same regardless of whether a bond is being compressed or expanded. While this is a good approximation for small displacements, it is incorrect when the bond is greatly stretched since the molecule dissociates and the energy approaches a constant value. A better representation of the potential is the Morse potential:

$$V(r) = D_e\{1 - \exp[-a(r - r_0)]\}^2$$

where D_e is the dissociation energy of the bond. This potential is said to be *anharmonic* and leads to a second term in the expression for the vibrational energy levels that depends on the *anharmonicity constant*, x_e:

$$u_n^v = \left(n + \frac{1}{2}\right)h\omega - \left(n + \frac{1}{2}\right)^2 x_e h\omega$$

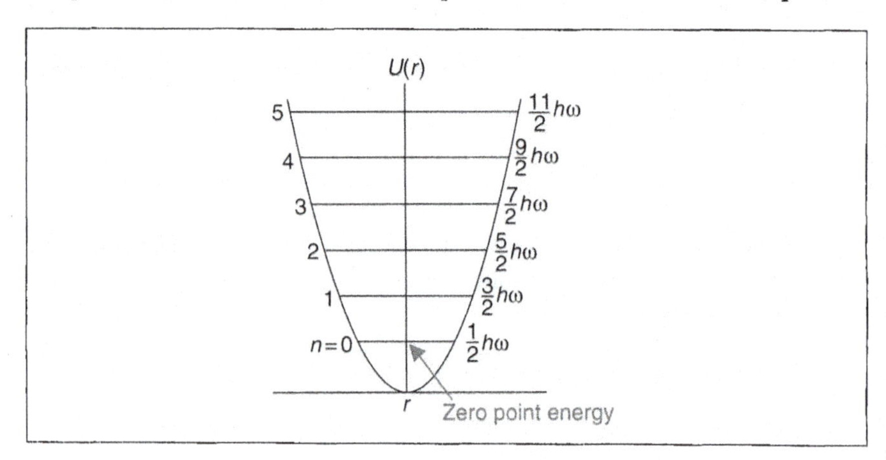

Figure 12.4 The vibrational energy levels for the harmonic oscillator superimposed on the associated potential energy curve

simultaneously. Hence the zero point motion prevents the molecule from ever being fully at rest, which would violate the above hypothesis.

To determine the molecular vibrational partition function, all we need to do is substitute in the expression for the energy levels (equation 12.22) to arrive at:

$$q^{V} = \sum_{n=0}^{\infty} \exp\left(-\frac{\left(n + \frac{1}{2}\right)h\omega}{k_B T} \right) \tag{12.23}$$

Unlike the other forms of energy we have considered so far, the spacing of the vibrational energy levels is almost always large compared to $k_B T$, at least under ambient conditions. Consequently the partition function must be treated as a *sum* rather than being approximated as an *integral* as we have done previously.

As with rotation, it turns out we can group together a number of terms that collectively have the units of Kelvin to yield a quantity known as the vibrational temperature, θ_v:

$$\theta_v = \frac{h\omega}{k_B} \tag{12.24}$$

The vibrational temperature is a measure of the onset of classical behaviour as the temperature increases, in the same sense as the rotational temperature introduced in equation (12.9). In Table 12.2 some examples of typical vibrational temperatures are given. For more complex molecules than diatomics there will be a different value for each mode due to the frequency of vibration varying. It is clear from the values in Table 12.2 that nearly all simple molecules will show quantization in their vibrational properties.

In this text, the unit used for vibrational frequencies is **wavenumbers**, cm^{-1}. Although this is not an SI unit, it is still the most widely employed unit in the literature, and therefore it is important to be familiar with handling wavenumbers. To convert a vibrational frequency in wavenumbers to one in the SI unit of Hz (s^{-1}) it is necessary to multiply by the *speed of light* expressed as 2.998×10^{10} cm s^{-1}.

Table 12.2 Vibrational temperatures for selected molecules. Note that for polyatomic molecules the vibrational temperatures for each of the normal modes are given

Molecule		θ_v/K	*Molecule*		θ_v/K
H_2		6210	CO		3070
N_2		3340	NO		2690
O_2		2230	HCl		4140
Cl_2		810	HBr		3700
Br_2		470	HI		3200
CO_2	ω_1	1890	H_2O	ω_1	5410
	ω_2	3360		ω_2	5250
	$\omega_3 = \omega_4$	954		ω_3	2290

It is possible to develop further the expression for the vibrational partition function, even though it must remain as a summation. Writing out the sum:

$$q^{\mathrm{V}} = \exp\left(-\frac{\theta_{\mathrm{v}}}{2T}\right) + \exp\left(-\frac{3\theta_{\mathrm{v}}}{2T}\right) + \exp\left(-\frac{5\theta_{\mathrm{v}}}{2T}\right) + \ldots \qquad (12.25)$$

We can factor out the term arising from the zero point energy:

$$q^{\mathrm{V}} = \exp\left(-\frac{\theta_{\mathrm{v}}}{2T}\right)\left[1 + \exp\left(-\frac{\theta_{\mathrm{v}}}{T}\right) + \exp\left(-\frac{2\theta_{\mathrm{v}}}{T}\right) + \ldots\right] \qquad (12.26)$$

To clarify the following working, let us make the substitution:

$$x = \exp\left(-\frac{\theta_{\mathrm{v}}}{T}\right) \qquad (12.27)$$

which leads to:

$$q^{\mathrm{V}} = x^{1/2}\left(1 + x + x^2 + x^3 + x^4 + \ldots\right) \qquad (12.28)$$

Box 12.1 Mathematical Aside

The following series is a geometric progression:

$$S = 1 + x + x^2 + x^3 + x^4 + \ldots \qquad (12.29)$$

We can obtain the sum of the series, provided $x < 1$, by multiplying each term in the above equation by x:

$$xS = x + x^2 + x^3 + x^4 + x^5 + \ldots \qquad (12.30)$$

and then by subtracting equation (12.30) from equation (12.29) to obtain:

$$(1 - x)S = 1 \qquad (12.31)$$

This result is arrived at because x raised to a very large power is negligibly small, since we have already said that x must be less than one, and therefore the terms at the end of the series are not important. Rearranging the above to make S the subject of the formula we obtain the final result:

$$S = \frac{1}{1 - x} \qquad (12.32)$$

Identifying the term in parentheses in equation (12.28) as just a standard geometric progression allows us to use the result for the sum of such a series given in equation (12.32). Note that because the vibrational frequency and temperature are always positive, we can demonstrate that x is necessarily less than 1 and therefore the sum will be convergent. Hence, we can now eliminate the explicit summation from the vibrational partition function:

$$q^{\mathrm{V}} = \frac{x^{1/2}}{(1-x)} \tag{12.33}$$

Substituting back in for x using equation (12.27) we obtain:

$$q^{\mathrm{V}} = \frac{\exp(-\theta_{\mathrm{v}}/2T)}{1-\exp(-\theta_{\mathrm{v}}/T)} = \frac{\exp(\theta_{\mathrm{v}}/2T)}{\exp(\theta_{\mathrm{v}}/T)-1} \tag{12.34}$$

Note that the second equivalent form of the result given above is obtained from the first by multiplying both the top and bottom of the equation by $\exp(\theta_{\mathrm{v}}/T)$.

Finally, we can also express the result for the molecular vibrational partition function in terms of the fundamental constants and the vibrational frequency, instead of the vibrational temperature:

$$q^{\mathrm{V}} = \frac{\exp(-h\omega/2k_{\mathrm{B}}T)}{1-\exp(-h\omega/k_{\mathrm{B}}T)} \tag{12.35}$$

Worked Problem 12.3

Q What is the limit of the molecular vibrational partition function as the temperature goes to zero?

A As $T \to 0$ then all the exponential terms tend to $\exp(-\infty)$, which is equal to zero. Therefore for from equation (12.35) we obtain:

$$q^{\mathrm{V}} \to \frac{\exp(-\infty)}{1-\exp(-\infty)} = \frac{0}{1-0} = 0$$

Comment: this result may initially seem surprising owing to the interpretation of the partition function as the number of thermally accessible levels at a given temperature. When T goes to absolute zero we know that the molecule will be in the vibrational ground state. As this is singly degenerate, then the partition function ought to be 1 when $T = 0$ according to the physical interpretation. This would be true for the vibrational partition function if the energy was taken relative to the zero point value. However, since it is taken relative to the minimum on the harmonic curve, there is an extra multiplicative factor of $\exp(-\hbar\omega/2k_BT)$ in the partition function that tends to zero as the temperature approaches absolute zero owing to the zero point vibrational energy.

12.5 Thermodynamic Properties of the Harmonic Oscillator

As before with other forms of motion, we can now determine the vibrational contribution to the thermodynamic properties of a diatomic molecule that is behaving as a harmonic oscillator. However, unlike the previous cases we have considered, we cannot expect to arrive at simple classical results, because we have explicitly considered the quantization of vibration throughout the derivation of the partition function.

For a collection of N independent vibrating diatomic molecules with a single characteristic vibrational frequency, we can write down the total vibrational partition function by combining equations (11.9) and (12.34):

$$Q = (q^{\mathrm{V}})^N = \left(\frac{\exp(-\theta_{\mathrm{v}}/2T)}{1-\exp(-\theta_{\mathrm{v}}/T)}\right)^N \tag{12.36}$$

Substituting this into the expression for the energy in terms of the total partition function (10.12) and using the properties of logarithms to convert the power of N into a multiplier:

$$U^{\mathrm{V}} = Nk_BT^2\frac{\partial}{\partial T}\left(\ln\left(\frac{\exp(-\theta_{\mathrm{v}}/2T)}{1-\exp(-\theta_{\mathrm{v}}/T)}\right)\right) \tag{12.37}$$

This differentiation is most easily performed by using the results $\ln(A/B) = \ln(A) - \ln(B)$ and $\ln(\exp(A)) = A$. Thus:

$$U^{\mathrm{V}} = Nk_BT^2\frac{\partial}{\partial T}\left(-\frac{\theta_{\mathrm{v}}}{2T} - \ln(1-\exp(-\theta_{\mathrm{v}}/T))\right) \tag{12.38}$$

Performing the differentiation of the two terms leads to:

$$U^{\mathrm{V}} = \frac{Nk_B\theta_{\mathrm{v}}}{2} + Nk_B\frac{\theta_{\mathrm{v}}\exp(-\theta_{\mathrm{v}}/T)}{(1-\exp(-\theta_{\mathrm{v}}/T))} \tag{12.39}$$

$$U^{\mathrm{V}} = \frac{Nh\omega}{2} + Nk_{\mathrm{B}}\frac{\theta_{\mathrm{v}}}{\left(\exp(\theta_{\mathrm{v}}/T) - 1\right)} \tag{12.40}$$

The first term in the above expression is just the zero point energy of N harmonic oscillators, while the second term is the additional energy that arises from the excited vibrational states that are occupied at a particular temperature.

Worked Problem 12.4

Q Show that in the limit of high temperature, $T >> \theta_{\mathrm{v}}$, the molar vibrational energy relative to the zero point energy reduces to the classical limit of RT.

A When the temperature becomes very much larger than the vibrational temperature, it is possible to approximate the exponential in equation (12.40) by a series expansion:

$$\exp(-x) \approx 1 - x + \frac{x^2}{2!} - \frac{x^3}{3!} + \ldots \tag{12.41}$$

which is valid as x tends to zero. Since we are interested in taking the limit in which T tends to infinity, it is sufficient to only take the first two terms in the series in equation (12.41). Substituting this into equation (12.40) and neglecting the zero point energy term, since the energy relative to this is required, gives:

$$U^{\mathrm{V}} = N_{\mathrm{A}}k_{\mathrm{B}}\frac{\theta_{\mathrm{v}}}{\left(1 + \frac{\theta_{\mathrm{v}}}{T} - 1\right)} \tag{12.42}$$

$$U^{\mathrm{V}} = N_{\mathrm{A}}k_{\mathrm{B}}T = RT \tag{12.43}$$

Thus we have arrived at the classical result. Although we have previously said that the equipartition theorem states that each degree of freedom has $1/2k_{\mathrm{B}}T$ of energy, it turns out that each vibrational mode has twice this in energy. We can think of the system as having two degrees of freedom: one for the kinetic energy and one for the potential energy.

We can now further determine the molar heat capacity at constant volume for the harmonic oscillator through the relationship:

$$C_V = \left(\frac{\partial U_{\mathrm{m}}}{\partial T}\right)_V \tag{12.44}$$

Hence we obtain the following expression by substituting in equation (12.40) for 1 mole of atoms:

$$C_V = \left(\frac{\partial}{\partial T}\left(\frac{N_A h\omega}{2} + R\frac{\theta_v}{\left(\exp(\theta_v/T)-1\right)}\right)\right)_V \tag{12.45}$$

Performing the differentiation, the zero point energy disappears, since it is independent of temperature, to leave only the contribution of the second term:

$$C_V = R\left(\frac{\theta_v}{T}\right)^2 \frac{\exp(\theta_v/T)}{\left(\exp(\theta_v/T)-1\right)^2} \tag{12.46}$$

If we consider the limit of the molar heat capacity at high temperatures, *i.e.* when $T >> \theta_v$, we can introduce the same approximation for the exponential terms used in the previous worked problem:

$$C_V \approx R\left(\frac{\theta_v}{T}\right)^2 \frac{\left(1+\theta_v/T\right)}{\left(1+\theta_v/T-1\right)^2} = R\left(1+\frac{\theta_v}{T}\right) \tag{12.47}$$

Therefore the vibrational component of the molar heat capacity tends to R as the temperature goes to infinity. Note that we could also have arrived at the same result via the limiting behaviour of the internal energy at high temperature, as given in equation (12.43).

The characteristic shape of the temperature dependence of the molar heat capacity at constant volume is shown in Figure 12.5. We have already explained the fact that the curve reaches a plateau at high temperatures since the influence of quantization ceases to be important. As the temperature approaches absolute zero, the heat capacity also tends to zero. This is a direct consequence of quantization, since until a minimum level of thermal energy is achieved – such that excitation to the second vibrational level can occur – it will be impossible for the harmonic oscillator to adsorb any internal energy.

There are many other thermodynamic properties of a vibrating diatomic molecule that could be explored using the relationships

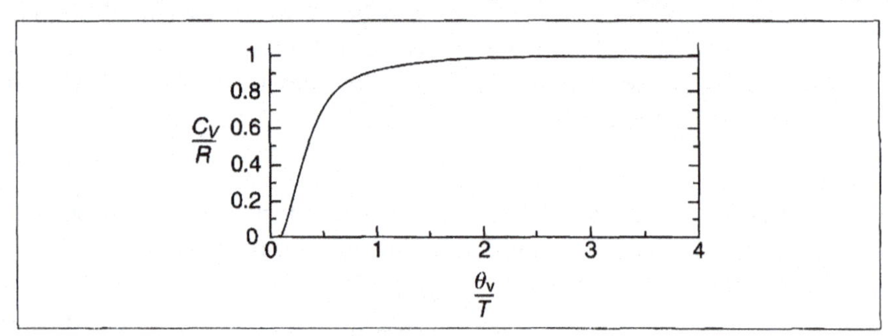

Figure 12.5 The molar heat capacity of vibration at constant volume as a function of temperature expressed as a ratio to the characteristic vibrational temperature of the mode

previously given. The approach to deriving these expressions and their characteristics is very similar to those illustrated above for the internal energy and heat capacity. Hence no further discussion will be given here, but some of these properties are the subject of the problems at the end of the chapter.

12.6 The Electronic Partition Function

Unlike the other forms of motion considered so far, it is impossible to write down a simple expression for the electronic energy levels. These must be obtained by solving the electronic Schrödinger equation as determined from quantum mechanics, and then used in the general expression for the partition function:

$$q^{\mathrm{E}} = \sum_{i=1}^{\text{all states}} g_i \exp\left(-\frac{u_i^{\mathrm{E}}}{k_{\mathrm{B}}T}\right) \qquad (12.48)$$

Note that it is important to remember that many electronic states have *degeneracies*, g_i, which must be taken into account.

Worked Problem 12.5

Q The energy required to excite an electron to the lowest unoccupied state of the H_2 molecule is approximately 1.6×10^{-18} J. Calculate the electronic partition function for molecular hydrogen at 1000 K, given that both the ground state and first excited state are singly degenerate, and assuming that no other states are important.

A In answering this question we can take the energies to be relative to the ground state (*i.e.* $u_0 = 0$). We just have to substitute into equation (12.48) for two terms using the energies given and taking g to be equal to 1 for both states:

$$q^{\mathrm{E}} = \exp\left(-\frac{0.0}{1.38066\times10^{-20}\ \mathrm{J}}\right) + \exp\left(-\frac{1.6\times10^{-18}\ \mathrm{J}}{1.38066\times10^{-20}\ \mathrm{J}}\right)$$
$$= \exp(0.0) + \exp(-115.89) \approx 1$$

Comment: the first excited state for hydrogen is so far above the ground state even at 1000 K that the second term is negligible.

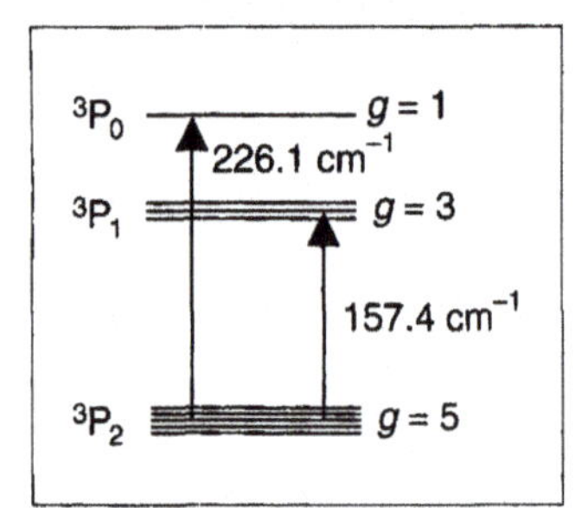

Figure 12.6 The first three electronic states of the oxygen atom

Because the separation between electronic states is, on the whole, very much larger than thermal energies, often only the ground state is important in the calculation of the partition function. Hence the electronic partition function is usually just equal to the degeneracy of the electronic ground state. Nearly all stable molecules have closed-shell singlet ground states, and hence the degeneracy, and therefore the partition function, is just equal to 1.

There are some important diatomic molecules that have open-shell ground states, such as O_2 and NO. Furthermore, many atoms have unpaired electrons when in the gas phase. For example, consider the oxygen atom whose low-lying electronic states are illustrated in Figure 12.6.

For an atom, the J quantum number for the total angular momentum can take values from $|L + S|$ to $|L - S|$, with a degeneracy of $2J + 1$. Hence the degeneracies of the $^3P_2$, $^3P_1$ and $^3P_0$ electronic states are 5, 3 and 1, respectively. Hence we can write down the electronic partition function as being:

$$q_0^{E} = 5\exp\left(-\frac{0}{k_BT}\right) + 3\exp\left(-\frac{157.4hc}{k_BT}\right) + \exp\left(-\frac{226.1hc}{k_BT}\right) \quad (12.49)$$

Remember that the *speed of light*, c, should be expressed in cm s^{-1}, given that the excitation energies are expressed in wavenumbers. At 300 K, for example, this would yield:

$$q_0^{E} = 5\exp(0) + 3\exp(-0.0075) + \exp(-0.0108) = 8.97 \quad (12.50)$$

The state of an electron is quantized and is determined by four **quantum numbers**:
n = principal quantum number (any positive integer, *i.e.* 1, 2, 3...)
l = orbital angular momentum (integer between 0 and $n - 1$)
m = orbital angular momentum in z-direction (integer between $+l$ and $-l$)
s = electron spin ($+\frac{1}{2}$ or $-\frac{1}{2}$)

For many electron systems, the individual orbital angular momentum contributions, l, combine to give the total orbital angular momentum, L, while the electron spin components similarly sum together to give S. In turn, the orbital and electron spin contributions sum together to the yield the total angular momentum of the system, J. Because this is a sum of vectors with several allowed orientations, more than one value of J can arise from a given L and S.

The state of a many-electron system can be described by a *term symbol*, $^{2S+1}L_J$, where the value of L is represented by a letter, S (L = 0), P (L = 1), D (L = 2), *etc.*, for an atom (or a capital Greek letter for molecules).

Summary of Key Points

1. *The rotational partition function*
   For a diatomic molecule, the rotational partition function can be expressed in terms of the moment of inertia as:
   $$q^R = \frac{8\pi^2 I k_B T}{\sigma h^2}$$
2. *The symmetry number*
   Because not all rotational energy levels are accessible when molecules have symmetry, as a consequence of the Pauli exclusion principle, it is necessary to divide the rotational partition function by a symmetry number, σ. This corrects for the fraction of rotational levels that can be occupied. Homonuclear diatomic molecules have a symmetry number of 2, regardless of whether the nuclei are bosons or fermions, while for heteronuclear diatomics the value is one.

3. *The vibrational partition function*
   The vibrational partition function for a diatomic molecule with a single frequency, ω, is given by:

$$q^{\mathrm{v}} = \frac{\exp(-h\omega/2k_{\mathrm{B}}T)}{1-\exp(-h\omega/k_{\mathrm{B}}T)}$$

4. *The rotational and vibrational temperatures*
   For both rotation and vibration the partition functions and thermodynamic properties can be expressed in terms of constants which have the units of temperature. When the actual temperature is much greater than the rotational temperature, θ_{r}, or the vibrational temperature, θ_{v}, then both types of motion show classical behaviour, whereas at temperatures below these they exhibit the influence of quantization.
5. *The electronic partition function*
   Because the electronic energy levels cannot be generalized, a sum over all states must be made, allowing for the degeneracy of each one. For most closed-shell molecules only the ground state is occupied, and therefore the electronic partition function is just equal to the degeneracy of this state.

Further Reading

P. W. Atkins, *Physical Chemistry*, 6th edn., Oxford University Press, Oxford, 1998, chapters 19 and 20.

C. E. Hecht, *Statistical Thermodynamics and Kinetic Theory*, Freeman, New York, 1990, chapter 2.

A. Maczek, *Statistical Thermodynamics*, Oxford University Press, Oxford, 1998.

D. A. McQuarrie and J. D. Simon, *Molecular Thermodynamics*, University Science Books, Sausalito, Calif., 1999.

Problems

1. Derive expressions for the rotational contribution to the following thermodynamic quantities for a diatomic molecule: (a) the pressure, (b) the entropy.

2. Calculate the rotational temperature for a molecule of Cl_2, given that the molar mass of Cl is 35.453 g mol^{-1} and the bond length is 0.1986 nm. What would be the rotational partition function for this molecule at a temperature of 298.15 K?

3. Calculate the vibrational contribution to the internal energy of a mole of dichlorine molecules at 298 K, given that the vibrational frequency is 561.1 cm^{-1}. How does this result compare with the classical limit?

4. Derive an expression for the vibrational entropy of a harmonic oscillator.

5. Calculate the electronic partition function for the OH radical at 298 K, given that there are two doublet electronic states separated by 139.7 cm^{-1}.

6. The molecule NO has a doubly degenerate ground state and a low-lying first excited state of the same multiplicity at a relative energy of 121.1 cm^{-1}. Assuming that no other electronic states contribute: (a) derive an expression for the heat capacity, C_V; (b) sketch the variation of the heat capacity relative to the Boltzmann constant with temperature; and (c) explain the shape of the heat capacity curve.

13
Statistical Mechanics and Equilibrium

The partition function has now been derived for the four principal forms of energy that we encounter in molecular systems, and we have examined how these results may be utilized to arrive at basic thermodynamic quantities associated with the physical state of a gas. In this chapter we will consider how statistical mechanics can be employed to determine the quantities that control chemical reactions in the gas phase, namely the *Gibbs free energy* and the *equilibrium constant*, K_p.

Aims

In this chapter we will examine how to calculate the thermodynamics of chemical reactions in the gas phase and the position of equilibrium. By the end of this chapter you should be able to:

- State orders of magnitude for the partition functions of different types of energy
- Express the Gibbs free energy in terms of the partition functions for the reactants and products
- Calculate the equilibrium constant for a gas phase reaction based on the partition functions of the chemical components

13.1 Thermodynamics of Gaseous Molecules

Earlier we separated the total energy of an ideal gas atom or molecule into components associated with types of motion, namely translation, rotation and vibration, plus the electronic contribution. Having examined each one in detail, we now need to concern ourselves with the overall thermodynamic properties of gaseous species and so we must recombine the results we now have. In Table 13.1 there is a summary of the expressions that we have found for the partition functions, along with a guide to typical orders of magnitude. The latter quantity is particularly useful, as it provides a useful check on whether any values we calculate are sensible. Also included are some generalizations of the formulae we have previously encountered for situations beyond the diatomic molecule case so far considered.

Table 13.1 Summary of the partition functions for an ideal gas. Here V represents the volume (in m^3), m represents the mass of a molecule, I is the moment of inertia for a linear molecule, I_A, I_B and I_C are the moments of inertia for a non-linear molecule about the principal axes, and σ is the symmetry number.

Energy	*Degrees of freedom*	*Partition function*	*Order of magnitude*
Translational	3	$\left(\frac{2\pi m k_B T}{h^2}\right)^{3/2} V$	10^{31}–$10^{32} \times V$
Rotational (linear)	2	$\frac{8\pi^2 I k_B T}{\sigma h^2}$	10–10^2
Rotational (non-linear)	3	$\frac{\left(I_A I_B I_C\right)^{1/2}}{\sigma}\left(\frac{8\pi^2 k_B T}{h^2}\right)^{3/2}$	10^2–10^3
Vibration (per mode)	1	$\frac{\exp\left(-h\omega/2k_B T\right)}{\left(1-\exp\left(-h\omega/k_B T\right)\right)}$	1–10

In the case of a polyatomic molecule with m normal modes of vibration (instead of just one as we have so far encountered), each with a harmonic frequency of ω^m, the molecular vibrational partition function will just be the product of the values for each mode:

$$q^V = \prod_m \left[\frac{\exp\left(-h\omega^m/2k_B T\right)}{1-\exp\left(-h\omega^m/k_B T\right)}\right] \tag{13.1}$$

Although we have been concerned only with gases so far, and have restricted ourselves to ideal gases for the calculation of individual partition functions, the concepts we have developed can be applied more generally. For example, it is equally possible to use the foregoing results to study solids as well. In this case it is only necessary to consider vibrational motion. Hence the partition function for a solid may be determined using equation (13.1), where the product is of the terms for all of the phonons, which are the vibrational modes of the material.

13.2 The Gibbs Free Energy

In the previous chapters it has been demonstrated how it is possible to calculate the Helmholtz free energy on a statistical mechanical basis.

However, most gas phase reactions occur under conditions of constant temperature and pressure, where the appropriate thermodynamic quantity is the Gibbs free energy. Our aim now is to show how this may also be arrived at.

Using the expressions of classical thermodynamics already previously introduced in Chapter 5, specifically combining the equations (5.4) and (5.5) with equation (3.6), the relationship between the Gibbs and Helmholtz free energies can be seen to be:

$$G = A + pV \tag{13.2}$$

If we continue to consider only the case of an ideal gas of n moles, the second term above can be replaced by using the ideal gas equation:

$$G = A + nRT \tag{13.3}$$

The Helmholtz free energy can now be expressed in terms of the total partition function:

$$G - G_0 = -k_B T \ln Q + nRT \tag{13.4}$$

Note that we have replaced A_0 with G_0, since at absolute zero both terms are the same and equivalent to the zero point internal energy. Assuming we have N gaseous species present, then we may express the total partition function in terms of the molecular partition function, remembering to include the factor of $N!$ since this is for a gas:

$$G - G_0 = -k_B T \ln\left(\frac{q^N}{N!}\right) + nRT \tag{13.5}$$

$$G - G_0 = -Nk_B T \ln q + k_B T \ln N! + nRT \tag{13.6}$$

Again it is possible to use *Stirling's approximation* to eliminate ln $N!$ for the case of large N, as well as recalling that $Nk_B = nR$:

$$G - G_0 = -nRT \ln q + nRT \ln N - nRT + nRT \tag{13.7}$$

Simplifying and combining the logarithms gives:

$$G - G_0 = -nRT \ln\left(\frac{q}{N}\right) \tag{13.8}$$

The ratio q/N is actually an intensive quantity, *i.e.* it is independent of the number of moles. It is possible to demonstrate this by considering q

as the translational component of the partition function. For this case:

$$\frac{q}{N} = \left(\frac{2\pi mk_BT}{h^2}\right)^{3/2}\frac{V}{N} = \left(\frac{2\pi mk_BT}{h^2}\right)^{3/2}\frac{Nk_BT}{Np} \tag{13.9}$$

Here we have substituted for the volume, assuming that the ideal gas equation applies. This leaves an expression in which the number of particles cancels out, therefore proving that the ratio is intensive. We can now re-write equation (13.8) in terms of the molar partition function, q_m, and the Avogadro constant, N_A:

$$G - G_0 = -nRT\ln\left(\frac{q_m}{N_A}\right) \tag{13.10}$$

Note that the molar partition function here is the partition function for one atom or molecule moving around in the molar volume at a pressure p, and is not to be confused with the total molar partition function, which is that for one mole of species under the same conditions. The above arguments are not altered by the contributions of the other partition functions, since they are all already per molecule.

13.3 The Standard Gibbs Free Energy

The standard molar Gibbs function for substance J is defined at the standard pressure, $p^\ominus = 1$ bar, as:

$$G_J^\ominus - G_J^\ominus(0) = -RT\ln\left(\frac{q_J^\ominus}{N_A}\right) \tag{13.11}$$

where $q_J^\ominus$ is the partition function for one molecule of J in a volume $RT/p^\ominus$.

Let us return to the standard Gibbs function at zero temperature. For an ideal gas:

$$G = H - TS \tag{13.12}$$

$$G = U + nRT - TS \tag{13.13}$$

Hence when $T = 0$, $G = U$. Thus in the standard state:

$$G_J^\ominus = U_J^\ominus(0) - RT\ln\left(\frac{q_J^\ominus}{N_A}\right) \tag{13.14}$$

where the first term on the right-hand side is now the molar internal energy of the substance J at absolute zero. This is the intrinsic energy in the atom or molecule.

Now consider a simple chemical reaction:

$$cC \rightarrow dD \tag{13.15}$$

in which c moles of a molecule C are transformed into d moles of a molecule D. Then the standard Gibbs free energy for the reaction will be:

$$\Delta_r G^\ominus = dG^\ominus_D - cG^\ominus_C \tag{13.16}$$

$$\Delta_r G^\ominus = dU^\ominus_D(0) - dRT\ln\left(\frac{q^\ominus_D}{N_A}\right) - cU^\ominus_C(0) + cRT\ln\left(\frac{q^\ominus_C}{N_A}\right) \tag{13.17}$$

$$\Delta_r G^\ominus = \Delta_r U^\ominus(0) - RT\ln\left[\frac{\left(q^\ominus_D/N_A\right)^d}{\left(q^\ominus_C/N_A\right)^c}\right] \tag{13.18}$$

$\Delta_r U^\ominus(0)$ is the difference between the intrinsic internal energies of the products and reactants, as illustrated in Figure 13.1. The second term takes into account the fact that at non-zero temperatures the reactants and products can take up the thermal energy in different ways.

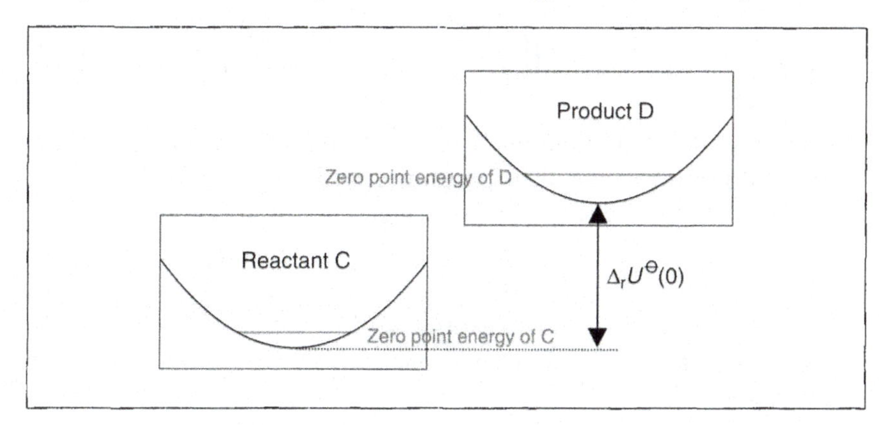

Figure 13.1 The difference between the intrinsic energies of the products and reactants. Note the energies are measured from the bottom of the vibrational well and not the zero point energy

We can readily generalize equation (13.18) to any reaction with multiple reactants and products as follows:

$$\Delta_r G^\ominus = \Delta_r U^\ominus(0) - RT\ln\left[\prod_J \left(\frac{q^\ominus_J}{N_A}\right)^{\upsilon_J}\right] \tag{13.19}$$

where the product is over all the species in the reaction. The index υ_J is the number of moles associated with each species in the reaction. It takes a positive sign for products and a negative sign for reactants.

13.4 The Equilibrium Constant, K_p

Earlier, in equation (7.6), it was shown that the equilibrium constant for a gas phase reaction, K_p, is determined simply by the Gibbs free energy change associated with that reaction and the temperature:

$$\Delta_r G^\ominus = -RT \ln K_p \tag{13.20}$$

Combining equations (13.19) and (13.20) and dividing through by RT we obtain:

$$\ln K_p = -\frac{\Delta_r U^\ominus(0)}{RT} + \ln\left[\prod_J \left(\frac{q_J^\ominus}{N_A}\right)^{\nu_J}\right] \tag{13.21}$$

Using $A = \ln(\exp(A))$ for the first term on the right-hand side of equation (13.21):

$$\ln K_p = \ln\left(\left[\prod_J \left(\frac{q_J^\ominus}{N_A}\right)^{\nu_J}\right] \times \exp\left(-\frac{\Delta_r U^\ominus(0)}{RT}\right)\right) \tag{13.22}$$

This then simplifies to the final expression for the equilibrium constant:

$$K_p = \prod_J \left(\frac{q_J^\ominus}{N_A}\right)^{\nu_J} \times \exp\left(-\frac{\Delta_r U^\ominus(0)}{RT}\right) \tag{13.23}$$

Hence if we know the internal energy difference between a set of reactants and products, plus their partition functions, it is possible to calculate the position of chemical equilibrium.

Worked Problem 13.1

Q Calculate the equilibrium constant for the isotopic exchange reaction:

$$H_{2(g)} + D_{2(g)} \rightleftharpoons 2HD_{(g)}$$

at 800 K and a pressure of 200,000 Pa, given the following data:

H_2: $B = 60.864$ cm^{-1}, $\omega = 4400.39$ cm^{-1}, $D_0 = 432.06$ kJ mol^{-1}
D_2: $B = 30.442$ cm^{-1}, $\omega = 3118.46$ cm^{-1}, $D_0 = 439.60$ kJ mol^{-1}
HD: $B = 45.638$ cm^{-1}, $\omega = 3811.46$ cm^{-1}, $D_0 = 435.50$ kJ mol^{-1}

The molar masses of H and D are 1.008 and 2.014 g mol^{-1}, respectively.

A The expression for the equilibrium constant in the case of this reaction is:

$$K_p = \left(\frac{(q_{HD}/N_A)^2}{(q_{H_2}/N_A)(q_{D_2}/N_A)}\right)\exp\left(-\frac{\Delta U(0)}{RT}\right)$$

For each molecule the molecular partition function is given by:

$$q = q^T q^R q^V$$

since the electronic partition will be simply 1 in each case. To calculate the component partition functions we can use the formulae from previous chapters, while expressing the rotational partition function in the simplified form $q^R = k_B T/\sigma B$ as the values of the rotational constants are given. Below we will give the working for H_2:

$$q^T_{H_2} = \left(\frac{2\pi m k_B T}{h^2}\right)^{3/2} \frac{RT}{p}$$

$$= \left(\frac{2\pi\left(\frac{2.016\times10^{-3}}{N_A}\right)1.38066\times10^{-23}\times800}{(6.6262\times10^{-34})^2}\right)^{3/2} \frac{8.314\times800}{2\times10^5}$$

$$= 4.0480\times10^{29}$$

$$q^R_{H_2} = \frac{1.38066\times10^{-23}\times800}{2\times60.864\times2.9979\times10^{10}\times6.6262\times10^{-34}} = 4.5678$$

$$\theta_v = \frac{hc\omega}{k_B} = \frac{6.6262\times10^{-34} - 2.9979\times10^{10}\times4400.39}{1.38066\times10^{-23}} = 6331.2 \text{ K}$$

$$q^V_{H_2} = \frac{\exp(\theta_V/2T)}{(\exp(\theta_V/T)-1)} = 1.913\times10^{-2}$$

Repeating for the other two molecules gives the results shown in Table 13.2.

Table 13.2 Summary of partition function contributions for D_2 and HD

Molecule	q^T	q^R	θ_v/K	q^V
D_2	1.143×10^{30}	9.1325	4486.8	6.078×10^{-2}
HD	7.429×10^{29}	12.1834	5484.5	3.249×10^{-2}

The final task is to calculate the value of $\Delta_r U(0)$, which is equal to the change in the dissociation energies, corrected by the change in the zero point energies:

$$\Delta_r U(0) = \Delta D_0 + \Delta_r U_{ZPE}$$

$$= 432.06 + 439.60 - 2 \times 435.50 + \frac{N_A hc}{2 \times 1000}$$

$$(2 \times 3811.9 - 4400.4 - 3118.5)\ \text{kJ mol}^{-1}$$

$$= 0.66 + 0.628\ \text{kJ mol}^{-1} = 1.288\ \text{kJ mol}^{-1}$$

Combining all the terms together to obtain the equilibrium constant yields:

$$K_p = 3.8536 \times \exp\left(-\frac{1288}{8.314 \times 800}\right) = 3.18\ (\text{to 3 s.f.})$$

Statistical mechanics can be even more powerful than this, since it is possible to go beyond static equilibrium properties, and even to predict the rates of reactions or kinetics for cases where the activation energy is large compared with thermal energies. This application of statistical mechanics is known as transition state theory, and is beyond the scope of this introductory text. However, the basic concept involves using equation (13.23) to calculate the equilibrium constant between the reactants and a so-called activated complex that represents the transition state for the process.

Summary of Key Points

1. *The Gibbs free energy*
   Through calculating the partition functions for 1 mole of reactants and products under standard conditions, as well as knowing the intrinsic internal energy difference at absolute zero temperature, the standard Gibbs free energy of reaction may be determined in the gas phase.

2. *The equilibrium constant*
   From knowing the standard Gibbs free energy of reaction, the equilibrium constant, K_p, for gas phase reaction can be determined through statistical mechanics.

Further Reading

P. W. Atkins, *Physical Chemistry*, 6th edn., Oxford University Press, Oxford, 1998, chapters 19 and 20.
R. P. H. Gasser and W. G. Richards, *Entropy and Energy Levels*, Clarendon Press, Oxford, 1974. chapter 5.
J. H. Knox, *Molecular Thermodynamics*, Wiley, Chichester, 1978, chapters 11 and 12.
D. A. McQuarrie and J. D. Simon, *Molecular Thermodynamics*, University Science Books, Sausalito, Calif., 1999.
K. Stowe, *Introduction to Statistical Mechanics and Thermodynamics*, Wiley, New York, 1984, chapter 21.

Problems

1. Calculate the standard Gibbs free energy of reaction for the formation of gaseous HCl from H_2 and Cl_2, given the following data:

H_2: $r = 0.07417$ nm, $\omega = 4400.39$ cm^{-1}, $D_0 = 432.06$ kJ mol^{-1}
Cl_2: $r = 0.1986$ nm, $\omega = 561.1$ cm^{-1}, $D_0 = 238.80$ kJ mol^{-1}
HCl: $r = 0.12746$ nm, $\omega = 2989.6$ cm^{-1}, $D_0 = 427.43$ kJ mol^{-1}

The molar masses of H and Cl are 1.008 and 35.455 g mol^{-1}, respectively. Standard pressure is 101,325 Pa and standard temperature is 298.15 K.

2. Determine the equilibrium constant under standard conditions for the gas phase dissociation of the diatomic lithium molecule:

$$Li_{2(g)} \rightleftharpoons 2Li_{(g)}$$

given that the dissociation energy, bond length and vibrational frequency of Li_2 are 99.38 kJ mol^{-1}, 0.2673 nm and 353.59 cm^{-1}, respectively, and that the molar mass of lithium is 6.94 g mol^{-1}. You may assume that only the ground electronic state of each species contributes to the partition function.

Glossary

Activated complex: transition state configuration for a chemical reaction that is considered to be in equilibrium with the reactants.

Activation energy: minimum internal energy barrier that has to be overcome for a chemical reaction to occur.

Activity, a_i: effective concentration of species i.

Activity coefficient, γ_i: ratio of activity to mole fraction (solvent) or molality (solute) for species i.

Adiabatic process: one in which no heat flow occurs; the system is thermally isolated.

Amount of substance in moles (mol), n: the number of atoms or molecules divided by N_A.

Anharmonicity: the deviation of a vibrating molecule from harmonic behaviour, with the energy levels becoming closer together with increasing vibrational quantum number and ultimately leading to dissociation.

Avogadro's constant, N_A: 6.022×10^{23} mol^{-1}.

Bar: unit of pressure; 1 bar = 100 kPa (exactly).

Boiling (vaporization) point, T_{vap}: temperature at which the vapour pressure of a liquid equals the external pressure. The *normal* boiling point T_b is equal to T_{vap} at p = 1 atm; the *standard* boiling point is defined at $p = p^\ominus$ = 1 bar.

Boltzmann distribution: $n_i/n_j = \exp(-\Delta U_{ji}/k_B T)$. Determines the occupancy of particles within energy levels for a macroscopic system.

Boltzmann's constant, k_B: 1.38066×10^{-23} J K^{-1}. The constant of proportionality between the statistical entropy and the natural logarithm of the number of microstates.

Born–Haber cycle: a closed path of reaction enthalpies which allows any one of the enthalpy changes to be determined if the others are known.

Born interpretation: the product of a wavefunction and its complex conjugate is equal to the electron density.

Canonical ensemble: the NVT ensemble, in which the specified properties are the number of particles, the volume and the temperature.

Chemical equilibrium: extent of reaction which minimizes the total Gibbs free energy G of the system.

Chemical potential, $\mu_i = (\partial G/\partial n_i)_{p,T,n_j \neq n_i}$: change of G with the number of moles of species i.

Clapeyron equation: gives the slope of the boundary between two phases in a (p, T) phase diagram, for example, the pressure dependence of the boiling point.

Classical theory: one in which properties, such as energy, are continuous and not quantized.

Clausius–Clapeyron equation: gives the temperature dependence of the vapour pressure of a liquid.

Clausius inequality, $dS \geq dq/T$: The entropy change of the system is greater than (for a spontaneous process) or equal to (for a reversible process) the heat flow divided by the temperature of the system.

Closed system: no exchange of matter, but heat flow and work are permitted.

Complexion: one of the possible indistinguishable arrangements of a set of particles within a microstate.

Component: one of the single species present in the system.

Composition, x: mole fraction units.

Concentration, c: mol dm^{-3}.

Critical point: where coexisting phases become indistinguishable (*e.g.* liquid and vapour).

Dalton's law: the pressure exerted by a mixture of ideal gases is the sum of the pressures exerted by the individual gases occupying the same volume alone.

Distinguishable: particles have a different chemical composition or are fixed in space, and are therefore unable to exchange positions through translation motion.

Electrical charge, q: elementary charge e (1.6×10^{-19} C); the charge on an electron is $-e$.

Electrical work, w_{el}: $dw_{el} = \Phi dq$.

Electric potential, Φ: units: volts ($1 \text{ V} = 1 \text{ J C}^{-1}$).

Endothermic process: $\Delta H > 0$ (heat absorbed; system gets colder, if thermally isolated).

Ensemble: a set of sampled configurations of a system, in contact with a thermal bath, where each individual sample may have a different state while sharing three common macroscopic properties.

Ensemble average: the value of a property when averaged over all the states explored within an ensemble.

Enthalpy, H: $H = U + pV$; ΔH is the amount of heat transferred at constant pressure (assuming only pV work performed).

Entropy, S: $dS = dw_{rev}/T$.

Equation of state: defines the thermodynamic state of the system in terms of a small number of system variables (for example, for an ideal gas: $p = nRT/V$).

Equilibrium: state of system where no further net spontaneous change occurs.

Equilibrium constant, K: defined in terms of the concentrations (or activities) of the reactants and products at equilibrium; also related to the standard reaction Gibbs free energy by $\Delta_r G^{\ominus} = -RT \ln K$.

Equilibrium constant, K_p: defined in terms of the partial pressures of the gaseous reactants and products at equilibrium.

Equipartition theorem: a classical theorem that states that every degree of freedom for motion has an energy of $\frac{1}{2}k_B T$.

Ergodic hypothesis: the ensemble average of a property is identical to the time average.

Exact differential: one which on integration yields a value of the function which is independent of the path of integration (for example, a state function).

Exothermic process: $\Delta H < 0$ (heat released; system gets hotter, if thermally isolated).

Expansion work, w_{exp}: pV work; $dw_{exp} = -p_{ex}\, dV$.

Extensive property: property that depends on the amount (mass) of substance present (*e.g.* V, U, H, G).

Extent of reaction, ξ: fraction of 1 mol of reaction as written, proceeding from reactants to products.

First Law: for any closed system there exists a property, *energy*, which is conserved and can be transferred into or out of the system by either *work* or *heat flow*. The internal energy of an isolated system is constant.

Fluctuation property: a property that is determined by the distribution of a quantity about its mean value, and is therefore related to the degree of fluctuation.

Free energy: *see* Gibbs and Helmholtz free energy.

Fugacity, f: effective pressure of a real gas.

Fundamental equation of chemical thermodynamics: $dG = V\,dp - S\,dT + \Sigma\mu_i\, dn_i$.

Gibbs free energy, G: $G = H - TS$. Spontaneous processes at constant pressure tend to minimize G.

Gibbs–Helmholtz equation: temperature dependence of G, in terms of H.

Group theory: deals with the set of symmetry elements present in a molecule and its properties. Can be used to determine whether vibrations are infrared or Raman active according to symmetry.

Harmonic oscillator: a vibrating system where the potential energy is given by $\frac{1}{2}k(r - r_0)^2$.

Heat, q: the transfer of energy that results from temperature differences, or to maintain isothermal conditions; heat is a path function.

Heat capacity (isobaric), C_p: rate of change of enthalpy H with T at constant pressure p.

Heat capacity (isochoric), C_V: rate of change of internal energy U with T at constant volume V.

Heisenberg's uncertainty principle: a fundamental hypothesis of quantum mechanics that states that certain pairs of observable properties, such as position/momentum and energy/time, cannot be simultaneously determined with arbitrary precision.

Helmholtz free energy, A: $A = U - TS$. Spontaneous processes at constant volume tend to minimize A.

Hess's law: the overall reaction enthalpy is the sum of the reaction enthalpies of the individual reactions into which the reaction may be divided.

Homonuclear: all nuclei are identical (*i.e.* all are the same isotope of the same element).

Ideal gas: a gas where the constituent atoms or molecules do not interact except when they collide.

Ideal gas equation: $pV = nRT$. The equation of state for an ideal gas.

Indistinguishable: particles have identical chemical composition and are able to move freely, such that they can exchange positions through translational motion.

Inexact differential: one whose integral yields a value of the function which depends on the path of integration (*e.g.* a path function).

Intensive property: property which does not depend on the amount (mass) of substance present (*e.g.* p, T, H_m, μ).

Internal energy, U: the total energy of a system (electronic, kinetic, *etc.*); we can only measure changes in U: $\Delta U = q + w$; $dU = dq + dw$.

Irreversible process: one that occurs spontaneously. The overall entropy of the system plus the surroundings increases. Less heat is absorbed, and less work done, than for the corresponding reversible process.

Isobaric: constant pressure.

Isochoric: constant volume.

Isolated system: no exchange of energy (heat, work) or matter with the surroundings.

Isoplethic: constant composition.

Isothermal: constant temperature.

Isothermal/isobaric ensemble: the npT ensemble, in which the fixed properties are number of moles, pressure and temperature.

Kinetics: the study of the rate of chemical reactions.

Kirchhoff's law: gives the temperature dependence of enthalpy changes in terms of the heat capacities.

Le Chatelier's principle: systems at equilibrium respond to perturbations so as to tend to minimize the effect of the perturbation.

Macroscopic system: one with a very large number of molecules or atoms (significant fraction of N_A).

Mass (atomic or molecular): m.

Mass (molar), M: atomic or molecular weight; $M = mN_A$.

Melting (fusion) point, T_{fus}: transition temperature from solid to liquid.

Microcanonical ensemble: the NVU (NVE) ensemble, in which the specified properties are the number of particles, the volume and the internal energy.

Microscopic system: one with a small number of atoms or molecules.

Microstate: a configuration of distinguishable particles within a given state.

Molality, m: number of moles of solute per kg of solvent.

Molar concentration (previously denoted "molarity"), c: number of moles per unit volume (mol dm^{-3}).

Molecular partition function, q: the partition function calculated on the basis of the energy levels within a single molecule.

Moment of inertia, I: the sum over all atoms within a molecule of the mass of that atom multiplied by the square of the distance from the centre of mass. This acts as a measure of how much force (torque) is required to rotate a molecule about a given axis.

Normal mode: one of the vibrations in a polyatomic system when treated as a harmonic oscillator. More formally, one of the eigenvectors obtained by diagonalizing the force constant matrix weighted by the inverse square roots of the atomic masses.

Open system: matter exchange, heat flow and work are all permitted.

Particle in a box: quantum mechanical model for a particle moving in a region of fixed length with zero potential energy.

Partition function: $\sum_{i=1}^{\text{all levels}} \exp(-U_i/k_B T)$

Pascal, Pa: unit of pressure; 1 Pa = 1 N m^{-2}.

Path function: property which depends on how the state of the system was attained, for example, work performed or heat transferred.

Pauli exclusion principle: the wavefunction must be antisymmetric with respect to the interchange of two fermions (*e.g.* electrons) and symmetric for the interchange of two bosons.

Perfect gas: one which obeys the ideal gas law $pV = nRT$. The internal energy of the gas depends on T, but not p or V.

Phase: state of matter that is uniform in chemical composition and in physical state.

Phase diagram: plot showing regions of stability of phases as functions of two or more system variables (*e.g.* p, V, T, x_i).

Phase transition: transformation of one phase into another (*e.g.* melting of a solid to a liquid); the chemical potentials μ_i of all components present are equal in both phases at the transition point.

Phonon: vibrational mode of a solid.

Potential difference (EMF), *E*; units: volts ($1\ V = 1\ J\ C^{-1}$).

Pressure, *p*: force per unit area; units: pascal (Pa); $1\ Pa = 1\ N\ m^{-2}$.

***pV* work**: see expansion work.

Quantum mechanics: theory in which properties, such as energy, exist with discrete values, instead of as continuous variables.

Reaction Gibbs free energy, ΔG_r: slope of the Gibbs free energy with the extent of reaction, ξ.

Reduced mass, μ: for a diatomic molecule with two atoms of mass m_1 and m_2, this is equal to $(m_1 m_2 / m_1 + m_2)$.

Reversible process: one in which the system remains in equilibrium at each point; the change proceeds infinitely slowly and is able to do the maximum amount of work.

Rigid rotor: molecule whose bond lengths, bond angles and torsional angles remain fixed while rotating.

Rotational constant, *B*: determines the rotational energy of a rigid rotor according to $u_J^R = BJ(J + 1)$.

Rotational quantum number, *J*: specifies which rotational energy level a molecule occupies.

Rotational temperature, θ_r: a collection of fundamental constants (see equation 12.9), related to rotation, that determine approximately the temperature at which a rigid rotor makes the transition from quantum to classical behaviour.

Sacker–Tetrode equation: the expression for the translational contribution to the molar entropy of an ideal gas in terms of the pressure and temperature (see equation 11.39).

Schrödinger equation: $H\psi = U\psi$; equation that determines the wavefunction and energy levels for a time-independent system.

Second Law: the entropy of an isolated system increases during any spontaneous process.

Selection rule: states whether the transition between two different energy levels is allowed for forbidden.

Solute: the minor component in a solution.

Solution: a homogenous mixture of two or more components.

Solvent: the major component in a solution.

Spontaneous process: one in which the overall entropy (system plus surroundings) increases (*e.g.* any irreversible process); $(dG)_{T,p} \leq 0$; $(dA)_{T,V} \leq 0$; less heat is absorbed and less work done by the system than for the corresponding reversible process.

Standard ambient temperature: 298.15 K (25 °C).

Standard enthalpy change, $\Delta H^{\ominus}$: change in enthalpy per mole for a process (*e.g.* fusion, vaporization, solution, ionization), in which the initial and final species are in their standard states.

Standard enthalpy of formation, $\Delta_f H^{\ominus}$: the standard reaction enthalpy for the formation of a compound from its elements in their standard states, *e.g.* Ar(g), N_2(g), C(s, graphite).

Standard entropy (Third Law), $S^\ominus$: entropy at the specified temperature and $p = 1$ bar, relative to $S(T = 0) = 0$.

Standard Gibbs free energy of formation, $\Delta_f G^\ominus$: the standard reaction Gibbs free energy for the formation of a compound from its elements in their standard states.

Standard pressure, $p^\ominus$: $p = 1$ bar = 100 kPa (exactly); *note*: 1atm = 101.325 kPa = 760 Torr ≈ 760 mmHg.

Standard reaction enthalpy, $\Delta_r H^\ominus$: the change in enthalpy when the reactants change to products, both being in their standard states.

Standard reaction entropy, $\Delta_r S^\ominus$: the change in entropy when the reactants change to products, both being in their standard states.

Standard reaction Gibbs free energy, $\Delta_r G^\ominus$: the change in Gibbs free energy when the reactants change to products, both being in their standard states.

Standard state: the pure form of a substance at a specified temperature at 1 bar pressure ($p = p^\ominus$), denoted by a plimsoll ($^\ominus$) superscript.

State: the state of a system is defined by a small number of system variables (state functions), not all of which are independent (*e.g.* p, T, V, U, H, S, G).

State function: property of the system that is independent of how the sample was prepared.

Statistical mechanics: the theory that connects the microscopic level to macroscopic properties.

Stirling's approximation: $\ln N! \approx N \ln N - N$, for large N.

Stoichiometric coefficients, ν: in writing out a chemical reaction, the ν are the smallest integers consistent with the reaction, that is they are the smallest whole numbers of moles of each species involved in the reaction.

Surface work, w_γ: work of changing area σ due to surface tension γ; $dw_\gamma = \gamma\, d\sigma$.

Surroundings: region outside the system of interest, separated from it by a boundary.

Symmetry number, σ: the factor that the rotational partition function must be divided by to correct for the reduced number of accessible rotational states when symmetry is present.

System: region of interest, where the reaction or process takes place, *e.g.* a reaction vessel.

Thermal de Broglie wavelength: the associated quantum wavelength of a particle translating at a given temperature (see equation 11.22).

Thermal equilibrium: there is no tendency for heat flow; $T_{system} = T_{surroundings}$.

Thermochemistry: study of the heat produced or consumed by chemical reactions or processes.

Third Law: the entropy of all perfect materials is zero at $T = 0$ K, and is always positive for $T > 0$ K.

Total partition function, Q: the partition function for the complete system of particles. For N distinguishable particles this is related to the molecular partition function by $Q = q^N$, while for indistinguishable particles the relation is $Q = q^N/N!$

Transition state theory: statistical mechanics approach to calculating reaction rates. Assumes that there is an activated complex in equilibrium with the reactants, which goes through to the products with a frequency equal to that of the reactive vibrational mode.

Triple point: point where three phases coexist (*e.g.* ice, water and water vapour).

Trouton's rule: the entropy of vaporization $\Delta_{vap}S$ of many simple organic liquids is close to 85 J K^{-1} mol^{-1}.

Universal gas constant, R: $N_A k_B$ = 8.314 kJ K^{-1} mol^{-1}.

van't Hoff isochore: gives the temperature dependence of the equilibrium constant K in terms of $\Delta_r H^\ominus$.

Vapour pressure, p: partial pressure of a gas in equilibrium with its liquid.

Vibrational frequency, ω: the number of complete vibrational oscillations executed per second.

Vibrational temperature, θ_v: $h\omega/k_B$. This quantity approximately determines the temperature at which a harmonic oscillator makes the transition from quantum to classical behaviour.

Work, w: the change in energy that could be directly used to raise a weight somewhere in the surroundings; w is a path function.

Zero point energy: $\frac{1}{2}h\omega$; the minimum energy of any vibrational motion of frequency ω.

Zeroth Law: if A is in thermal equilibrium with B, and B is in thermal equilibrium with C, then C is also in thermal equilibrium with A.

Answers to Problems

Chapter 2

1. (i) The power supply performs an amount of electrical work on the system:

$$w_{elec} = Vit = \frac{V^2 t}{R}$$

Now $\Delta U = w + q$, and $q = 0$ for an adiabatic system. Thus:

$$\Delta U = w_{elec} = \frac{(20)^2 \times 50}{10} = +2 \text{ kJ}$$

Note that the electrical work is completely converted to heat in the heating element, raising the temperature of the system.
(ii) For the diathermic system, the heat flows out of the system into the thermal reservoir, cancelling out the work and leaving the system's internal energy unchanged:

$$\Delta U = \left(w_{elec} + q\right) = \left(+2 \text{ kJ} - 2 \text{ kJ}\right) = 0$$

2. (i) Substituting for $<v^2>$ gives:

$$U = \frac{3}{2}pV = \frac{3}{2}nRT$$

Thus U is directly proportional to temperature T.
(ii) At constant temperature, U is independent of both pressure p and volume V.

3. (i) If all the weights are removed at once, the external pressure p_{ex} drops suddenly from 10 atm to 1 atm, and the gas will expand irreversibly against the constant p_{ex} = 1 atm (due to the air in the room) until the internal pressure is also 1 atm. If the expansion is isothermal, its final volume will be 10 dm^3 and the work done is:

$$w = -(1 \text{ atm})\ (10 - 1)\ dm^3 = -9\ dm^3 \text{ atm} = -912 \text{ J}$$

Note that the work is negative, *i.e.* the work is done *by* the system. An amount of heat $q = -w$ will flow into the system to maintain $U = (q + w)$ = constant.

(ii) If the weights are removed in two stages, the work done is the sum of two irreversible steps, the first expansion occurring against a higher p_{ex} than the second. In the first stage, five weights are removed (p_{ex} = 5 atm) and the work done is:

$$w = -(5 \text{ atm})\ (2 - 1)\ dm^3 = -5\ dm^3 \text{ atm}$$

In the second stage, the remaining four weights are removed (p_{ex} = 1 atm) and:

$$w = -(1 \text{ atm})\ (10 - 2)\ dm^3 = -8\ dm^3 \text{ atm}$$

The total work is then: −13 dm^3 atm = −1318 J.

(iii) The same procedure is used to find that, if the weights are removed singly, the total work done is −1954 J.

If the expansion is carried out reversibly (via an infinite number of vanishingly small steps), with $p_{ex} = p$ at all stages, then:

$$w_{rev} = -nRT \ln\left(\frac{V_2}{V_1}\right)$$

Thus, for the expansion from (p_1 = 10 atm, V_1 = 1 dm^3) to (p_2 = 1 atm, V_2 = 10 dm^3), the reversible work is (with $nRT = pV$ = 10 dm^3 atm):

$$w_{rev} - (10 \ln 10)\ dm^3 \text{ atm} = -2333 \text{ J}$$

This is the *maximum* amount of work that can be done by the system.

4. Use $pV = nRT$ to give:

State	T/K	V/dm^3	p/Pa
1	273	22.4	1.013×10^5
2	546	22.4	2.026×10^5
3	546	44.8	1.013×10^5

Process A is isochoric ($dV = 0$ and therefore no work is done).
Process B is isothermal ($dV \neq 0$ and therefore work is done).
Process C is isobaric (constant pressure).
Given that $\Delta U = C_V \Delta T$ and $\Delta U = q + w$, we can write down expressions for ΔU, q and w:

Process	ΔU	q	w
A	$C_V\Delta T$	$C_V\Delta T$	0
B	0	$-w$	$-RT \ln (V_3/V_2)$
C	$C_V\Delta T$	$C_p\Delta T$	$C_V\Delta T - C_p\Delta T$

In order to calculate any values, we need to know C_V and C_p. For one mole of an ideal gas, $C_V = 3R/2 = 12.47$ J K^{-1} mol^{-1}, and $C_p = (C_V + R) = 5R/2 = 20.79$ J K^{-1} mol^{-1}. Then:

Process	ΔU/J	q/J	w/J
A	3404.6	3404.6	0
B	0	3146.5	–3146.5
C	–3404.6	–5674.3	2269.7
Total	0	876.8	–876.8

Intuitively, $\Delta U = 0$ for the complete cycle, since it is a state function.

Chapter 3

1.

$$U = \frac{3}{2}pV = \frac{3}{2}nRT$$

$U(298\text{ K}) = 3.72$ kJ mol^{-1}. Therefore:

$$H = U + pV = \frac{5}{2}pV = \frac{5}{2}nRT$$

$H(298\text{ K}) = 6.20$ kJ mol^{-1}

$$C_V = \left(\frac{\partial U}{\partial T}\right)_V = \frac{3}{2}nR$$

$C_{V,\text{m}} = 12.47$ J K^{-1} mol^{-1}

$$C_p = \left(\frac{\partial H}{\partial T}\right)_p = \frac{5}{2}nR$$

$C_{p,\text{m}} = 20.79$ J K^{-1} mol^{-1}

Note that C_V and C_p are independent of temperature for an ideal gas. Also note that $\left(\frac{\partial U}{\partial V}\right)_T = 0$ and $\left(\frac{\partial H}{\partial V}\right)_T = 0$, *i.e.* the internal energy, U, and the enthalpy, H, of an ideal gas are *independent of volume* at constant temperature.
The relationship between C_p and C_V may be derived by writing:

$$H = (U + pV) = (U + nRT)$$
$$\mathrm{d}H = \mathrm{d}U + nR\mathrm{d}T$$
$$(\mathrm{d}q)_p = (\mathrm{d}q)_V + nR\mathrm{d}T$$

Therefore, for an ideal gas:

$$C_p = C_V + nR$$

This agrees with the above expressions for C_p and C_V.

2. 100 g of KNO_3 (M_r = 101.11) = 0.989 mol

$$\Delta_{sol}H^{\ominus}(298\text{ K}) = +34.9\text{ kJ mol}^{-1}$$
$$C_{p,m}(H_2O) = 75.29\text{ J K}^{-1}\text{ mol}^{-1}$$
$$\Delta T = -\frac{\Delta H}{C_p} = -\frac{\Delta H}{55.5\ C_{p,m}} = -\frac{34{,}900 \times 0.989}{55.5 \times 75.29}$$
$$= -8.26\ °C$$
$$T_{final} = 16.7\ °C$$

100 g of $AlCl_3$ (M_r = 133.33) = 0.750 mol

$$\Delta_{sol}H^{\ominus}(298\text{ K}) = -329\text{ kJ mol}^{-1}$$
$$\Delta T = -\frac{(-329{,}000) \times 0.750}{55.5 \times 75.29}$$
$$= +59\ °C$$
$$T_{final} = 84\ °C$$

3. The integrated form of the Kirchhoff equation permits the calculation of the enthalpy of reaction at one temperature if it is known at a different temperature, together with the necessary heat capacity data:

$$\Delta H(T_2) = \Delta H(T_1) + \int_{T_1}^{T_2} \Delta C_p \, dT$$

Hence for this particular problem:

$$\Delta H(T_2) = -46{,}110 + \int_{298}^{700} (\sum va + \sum vbT + \sum vcT^{-2}) \, dT$$

Expanding the three summations leads to:

$$\sum va = a(NH_3) - \tfrac{1}{2}a(N_2) - \tfrac{3}{2}a(H_2)$$
$$= 29.75 - \frac{28.58}{2} - \frac{3 \times 27.28}{2}$$
$$= -25.46$$
$$\sum vb = 18.33 \times 10^{-3}$$
$$\sum vc = -2.05 \times 10^{5}$$

Therefore:

$$\Delta H(700\ \mathrm{K}) = -46{,}110 - 25.46[700 - 298]$$
$$+18.33 \times 10^{-3}\left[\frac{700^2 - 298^2}{2}\right] + 2.05 \times 10^{5}\left[\frac{1}{700} - \frac{1}{298}\right]$$
$$= -46{,}110 - 10{,}231 + 3676 - 395$$

Thus, $\Delta_r H^\circ(700\ \mathrm{K}) = -53.06\ \mathrm{kJ\ mol^{-1}}$

Chapter 4

1. (i) Freezing of a supercooled liquid is an irreversible change because, unlike at 273.15 K, the system does not remain at equilibrium during the freezing process, *i.e.*, it cannot be reversed by an infinitesimal change in one of the conditions (T, p).

(ii) This is done by devising a set of reversible steps, the sum of which is equivalent to the single irreversible step. Since state functions are independent of the path taken, the value calculated is the same in both cases.

(iii) For this problem we can devise the following reversible path between the initial and final states: (a) heat the water from 263.15 K to 273.15 K, at constant pressure p°; (b) freeze the water at 273.15 K to ice at 273.15 K at p°. This is a phase transition under equilibrium conditions and hence is a reversible process; (c) cool the ice from 273.15 K to 263.15 K at constant pressure p°. For a change in temperature at constant pressure, the entropy change is given by:

$$\Delta S = \int_{T_1}^{T_2} C_p \frac{dT}{T} = \int_{T_1}^{T_2} C_p \, d\ln T$$

The entropy of fusion is given by:

$$\Delta_{fus}S^{\Theta} = \Delta_{fus}H^{\Theta}/T_{fus}$$

Note that freezing at 273.15 K will involve an entropy change of $(-\Delta_{fus}S^{\Theta})$.
Thus, for the reversible steps, we can write:

$$\Delta S = \int_{263}^{273} C_p(\text{liq } H_2O)\frac{dT}{T} - \frac{\Delta_{fus}H}{T_{fus}} + \int_{273}^{263} C_p(\text{ice})\frac{dT}{T}$$

$$= 75.29\ln\frac{273.15}{263.15} - \frac{6008}{273.15} + 36.9\ln\frac{263.15}{273.15}$$

$$= 2.81 - 21.99 - 1.38$$

$\Delta_{fus}S^{\Theta} = -20.56$ J K^{-1} mol^{-1}
This is the decrease in entropy of the supercooled water on freezing to ice at 263.15 K. To find the entropy change of the surroundings, we can assume that the large heat reservoir of the surroundings remains at constant temperature during the essentially reversible transfer to it of $\Delta_{fus}H^{\Theta}$ at 263.15 K. We calculate $\Delta_{fus}H^{\Theta}$(263.15 K) from the Kirchhoff equation (since the pressure is constant at p^{Θ}):

$$d(\Delta_{fus}H^{\Theta})/dT = \Delta C_p = [C_p(\text{liq H2O}) - C_p(\text{ice})] = 38.39 \text{ J K}^{-1} \text{ mol}^{-1}$$

$$\Delta_{fus}H^{\Theta}(273.15 \text{ K}) - \Delta_{fus}H^{\Theta}(263.15 \text{ K}) = 38.39[273.15 - 263.15] = 383.9 \text{ J mol}^{-1}$$

Hence, $\Delta_{fus}H^{\Theta}$(263.15 K) = (6008 − 383.9) = +5.624 kJ mol^{-1}
and $\Delta S^{surr} = \Delta_{fus}S^{\Theta}$(263.15 K) = (5624/263.15) = 21.37 J K^{-1} mol^{-1}

The total entropy change is thus:

$$\Delta S^{total} = (\Delta S^{surr} + \Delta S) = (21.37 - 20.56) = +0.81 \text{ J K}^{-1} \text{ mol}^{-1}$$

The entropy of the universe has increased during the irreversible freezing process, as the Second Law dictates.

2. (i) The general expression we need is equation (4.13):

$$\Delta S = C_V \ln\left(\frac{T_B}{T_A}\right) + nR\ln\left(\frac{V_B}{V_A}\right)$$

Under isobaric (constant pressure) conditions, we can write:

$$\left(\frac{V_B}{V_A}\right) = \left(\frac{nRT_B/p}{nRT_A/p}\right) = \left(\frac{T_B}{T_A}\right)$$

Hence:

$$\Delta S = C_V \ln\left(\frac{T_B}{T_A}\right) + nR\ln\left(\frac{T_B}{T_A}\right)$$

$$= (C_V + nR)\ln\left(\frac{T_B}{T_A}\right)$$

$$= C_p \ln\left(\frac{T_B}{T_A}\right)$$

(ii) Under isothermal conditions, we can use equation (4.16):

$$\Delta S = nR\ln\left(\frac{V_B}{V_A}\right) = nR\ln\left(\frac{p_A}{p_B}\right) = -nR\ln\left(\frac{p_B}{p_A}\right)$$

Setting $S_B = S$, $S_A = S^{\ominus}$, $p_B = p$, $p_A = p^{\ominus}$ leads directly to:

$$S = S^{\ominus} - nR\ln\left(\frac{p}{p^{\ominus}}\right)$$

3.

$$\Delta S = \int_{T_1}^{T_2} \frac{C_p(T)}{T} \mathrm{d}T$$

$$= a\int \frac{\mathrm{d}T}{T} + b\int \mathrm{d}T + c\int \frac{\mathrm{d}T}{T^3}$$

$$= a\ln\left(\frac{T_2}{T_1}\right) + b(T_2 - T_1) - \frac{c}{2}\left[\frac{1}{T_2^2} - \frac{1}{T_1^2}\right]$$

$$\Delta S = 29.96\ln\left(\frac{423.15}{298.15}\right) + 4.18\times10^{-3}(423.15 - 298.15)$$

$$-\frac{-1.67\times10^5}{2}\left[\frac{1}{423.15^2} - \frac{1}{298.15^2}\right]$$

$$= 10.49 + 0.52 - 0.47$$

$$= +10.54 \text{ J K}^{-1} \text{ mol}^{-1}$$

Chapter 6

1. Integration of the Clapeyron equation, assuming that the enthalpy and volume changes upon melting are independent of temperature and pressure, gives:

$$p = p^* + \frac{\Delta_{\text{fus}}H}{\Delta_{\text{fus}}V}\ln\left(\frac{T}{T^*}\right)$$

For water:

$$\begin{aligned}\Delta_{\text{fus}}V &= V_{\text{m}}(\text{water}) - V_{\text{m}}(\text{ice})\\ &= 18.0 - 19.7\\ &= -1.7\ \text{cm}^3\ \text{mol}^{-1}\\ \Delta_{\text{fus}}H^{\text{o}} &= +6.008\ \text{kJ mol}^{-1}\end{aligned}$$

Thus:

$$\begin{aligned}p &= p^* - \frac{6008}{1.7\times10^{-6}}\ln\left(\frac{T}{T^*}\right)\\ &= 10^5 - 3.534\times10^9\ln\left(\frac{263.15}{273.15}\right)\\ &= 1319\times10^5\ \text{N m}^{-2}\\ &= 1302\ \text{atm}\ (1.32\ \text{kbar})\end{aligned}$$

2. The integrated form of the Clausius–Clapeyron equation, assuming $\Delta_{\text{vap}}H$ is independent of temperature, is:

$$\ln p = -\frac{\Delta_{\text{vap}}H}{RT} + \text{const.}$$

A plot of the data in the form ln p versus T^{-1} gives a straight line of gradient:

$$-\frac{\Delta_{\text{vap}}H}{R} = -2973\ \text{K}$$

$$\Delta_{\text{vap}}H = 24.72\ \text{kJ mol}^{-1}$$

The normal boiling point T_{b} is the temperature at which the vapour pressure p becomes equal to 1 atm = 760 mmHg. This value may be read off the plot, yielding T_{b} = 272.7 K (this is the same as the literature value of T_{b} = 272.7 K).

3. The standard Gibbs free energy of formation, $\Delta_{\text{f}}G^{\text{o}}$, is defined as the standard reaction Gibbs free energy, $\Delta_{\text{r}}G^{\text{o}}$, for formation of the substance from its elements in their standard states (where they have $\Delta_{\text{f}}G^{\text{o}} = 0$):

$$\Delta_r G^{\Theta} = \Delta_r H^{\Theta} - T\Delta_r S^{\Theta}$$

$$\Delta_r G^{\Theta} = \sum_{\text{prod}} \nu_{\text{prod}} \Delta_f G^{\Theta}_{\text{prod}} - \sum_{\text{react}} \nu_{\text{react}} \Delta_f G^{\Theta}_{\text{react}}$$

Thus we write the reaction *per mole of the compound* as:

$$H_2(g) + \frac{1}{2}O_2(g) \rightarrow H_2O(l)$$

For liquid H_2O at 25 °C:

$$\Delta_f H^{\Theta} = -285.8 \text{ kJ mol}^{-1}$$

$$\begin{aligned}
\Delta_r S^{\Theta} &= S^{\Theta}(H_2O,l) - \left\{S^{\Theta}(H_2,g) + \frac{1}{2}S^{\Theta}(O_2,g)\right\} \\
&= (69.9 - 130.7 - 102.5) \\
&= -163.3 \text{ J K}^{-1} \text{ mol}^{-1} \\
\Delta_f G^{\Theta} &= \Delta_f H^{\Theta} - T\Delta_r S^{\Theta} \\
&= -285.8 \times 10^3 - (298.15 \times (-163.3)) \\
&= -237.1 \text{ kJ mol}^{-1}
\end{aligned}$$

Note that $\Delta_f H^{\Theta}$ and $\Delta_f G^{\Theta}$ are zero for elements in their standard states (at any temperature), even though the S^{Θ} are not.
The entropy change of the system, $\Delta_r S^{\Theta}$, is negative, reflecting the increased order of liquid water compared to gaseous H_2 and O_2.
Although the large negative value for $\Delta_f G^{\Theta}$ tells us that the *equilibrium* lies very strongly to the right (*i.e.*, towards liquid water), it says nothing about the *rate* at which equilibrium will be attained. Indeed, a gaseous mixture of H_2 and O_2 may be kept for a very long time without reacting, owing to the high energy barrier between reactants and products.
The entropy change of the surroundings is given by

$$\begin{aligned}
\Delta S^{\text{surr}} &= -\frac{\Delta_r H^{\Theta}}{T} \\
&= -\frac{(-285.8 \times 10^3)}{298.15} \\
&= +958.6 \text{ J K}^{-1} \text{ mol}^{-1}
\end{aligned}$$

Thus the formation of liquid water from gaseous hydrogen and oxygen leads to a drop in entropy of the system, but a much larger increase in entropy of the surroundings, such that the overall entropy change is positive and the process is spontaneous (as seen directly by the negative sign of $\Delta_f G^{\Theta}$).

Chapter 7

1. This problem requires the use of:

$$\Delta_r G^\ominus = -RT \ln K_p$$

In this equation, $\Delta_r G^\ominus$ is the standard reaction Gibbs free energy at the standard pressure $p^\ominus$ (1 bar), and K_p is the (dimensionless) equilibrium constant, defined by:

$$K_p = \frac{(p_{CO}/p^\ominus)^2(p_{O_2}/p^\ominus)}{(p_{CO_2}/p^\ominus)^2} = \frac{p_{CO}{}^2 p_{O_2}}{p_{CO_2}{}^2 p^\ominus}$$

If the fractional dissociation of CO_2 is denoted by α, and the gas mixture is at a total pressure p_0, we have, starting with 2 moles of CO_2:

	CO_2	CO	O_2	Total
No. of moles	$2(1-\alpha)$	2α	α	$2+\alpha$
Partial pressures	$\frac{2(1-\alpha)p_0}{2+\alpha}$	$\frac{2\alpha p_0}{2+\alpha}$	$\frac{\alpha p_0}{2+\alpha}$	p_0

Therefore:

$$K_p = \frac{\alpha^3}{(1-\alpha)^2(2+\alpha)} \frac{p_0}{p^\ominus}$$

For this problem we know that $p_0 = p^\ominus = 1$ bar, and hence, neglecting α terms in the denominator (since $\alpha << 1$):

$$K_p(1400\ \text{K}) = \frac{(1.27 \times 10^{-4})^3}{2} = 1.024 \times 10^{-12}$$

The variation of K_p with temperature is given by the van't Hoff isochore:

$$\frac{\mathrm{d} \ln K_p}{\mathrm{d}(1/T)} = -\frac{\Delta_r H^\ominus}{R}$$

If $\Delta_r H^\ominus$ is assumed to be independent of temperature, we obtain for the two temperatures T_1 and T_2:

$$\ln K_p(T_2) - \ln K_p(T_1) = -\frac{\Delta_r H^\ominus}{R}\left(\frac{T_1 - T_2}{T_1 T_2}\right)$$

With T_1 = 1000 K and T_2 = 1400 K, this becomes:

$$\ln(1.024\times10^{-12})-\ln(4\times10^{-21})=-\frac{\Delta_r H^\Theta}{8.314}\left(\frac{-400}{1000\times1400}\right)$$

Therefore:

$$\Delta_r H^\Theta = 8.314 \times 3500 \times (-27.61 + 46.97) = +563.4 \text{ kJ mol}^{-1}$$
$$\Delta_r G^\Theta \text{ (1000 K)} = -8.314 \times 1000 \times \ln(4 \times 10^{-21}) = +390.5 \text{ kJ mol}^{-1}$$

We can then calculate the reaction entropy using:

$$\Delta_r S^\Theta = \frac{\Delta H^\Theta - \Delta G^\Theta}{T}$$

Thus:

$$\Delta_r S^\Theta(1000 \text{ K}) = +172.9 \text{ J K}^{-1} \text{ mol}^{-1}$$

Note that the standard reaction entropy is positive, since 2 moles of gas increase to 3 moles.

2.

$$\Delta_r G^\Theta = -RT \ln K_p = \Delta_r H^\Theta - T\Delta_r S^\Theta$$
$$\ln K_p = -(\Delta_r H^\Theta/R)(1/T) + (\Delta_r S^\Theta/R)$$

(i) Plot $\ln K_p$ versus $(1/T)$:
Gradient = $-(\Delta_r H^\Theta/R) = -21.76 \times 10^3$ K. Thus $\Delta_r H^\Theta = +181$ kJ mol^{-1}.
Intercept = $\Delta_r S^\Theta/R = 3.08$. Thus $\Delta_r S^\Theta = +25.6$ J K^{-1} mol^{-1}.
(ii) $\Delta_r G^\Theta$(1000 K) = $181 \times 10^3 - 1000(25.6) = +155.4$ kJ mol^{-1}. Thus $K_p = 7.63 \times 10^{-9}$.

$$K_p = \frac{(p_{NO}/p^\Theta)^2}{(p_{N_2}/p^\Theta)(p_{O_2}/p^\Theta)} = \frac{(p_{NO})^2}{(0.8)(0.2)}$$

Thus $p_{NO} = 3.49 \times 10^{-5}$ bar.
(iii) $\Delta_f G^\Theta \approx \Delta_r G^\Theta/2$, as expected from the definition of $\Delta_f G^\Theta$. In the calculation of $\Delta_r G^\Theta$, we neglected the temperature dependences of $\Delta_r H^\Theta$ and $\Delta_r S^\Theta$.

3. Define the degree of dissociation α, such that $\alpha = 0$ corresponds to pure A_2, and $\alpha = 1$ corresponds to pure A:

$$A_2(g) \rightleftharpoons 2A(g)$$
$$(1-\alpha) \qquad \alpha$$

The equilibrium constant is defined as:

$$K_p = \frac{\left(p_A / p^\ominus\right)^2}{p_{A_2} / p^\ominus}$$

where P_A and P_{A_2} are the partial pressures of A and A_2.
Now from Dalton's law:

$$p_A = x_A p$$
$$p_{A_2} = x_{A_2} p$$

where p is the total pressure ($p_A + p_{A_2}$).
We can thus express the mole fractions in terms of α:

$$x_A = \frac{2\alpha}{(1-\alpha)+2\alpha} = \frac{2\alpha}{1+\alpha}$$
$$x_{A_2} = \frac{(1-\alpha)}{(1-\alpha)+2\alpha} = \frac{1-\alpha}{1+\alpha}$$

Thus:

$$K_p = \frac{(x_A p)^2}{(x_{A_2} p)p^\ominus} = \frac{\left(\frac{2\alpha}{1-\alpha}\right)^2 p}{\left(\frac{1-\alpha}{1+\alpha}\right)p^\ominus} = \left(\frac{4\alpha^2}{1-\alpha^2}\right)\frac{p}{p^\ominus}$$

i.e.:

$$\alpha = \left(\frac{K_p}{K_p + \frac{4p}{p^\ominus}}\right)^{1/2}$$

Thus α decreases as p increases. When K_p is small compared with ($p/p^\ominus$), α is approximately proportional to $p^{-1/2}$.
For the dissociation of N_2O_4 to $2NO_2$ at 298 K:

$$\Delta_r G^\ominus = \Delta_r H^\ominus - T\Delta_r S^\ominus$$
$$= +57.2 \times 10^3 - 298 \times (+176)$$
$$= +4.73 \text{ kJ mol}^{-1}$$
$$K_p = \exp\left(-\frac{\Delta_r G^\ominus}{RT}\right) = 0.149$$

At a pressure $p = p^\ominus = 1$ bar:

$$\alpha = \left(\frac{0.1486}{0.1486 + 4}\right)^{1/2} = 0.189$$

At a pressure $p = 10p^{\ominus} = 10$ bar:

$$\alpha = \left(\frac{0.1486}{0.1486 + 40}\right)^{1/2} = 0.061$$

To calculate K_p at different temperatures, we assume that $\Delta_r H^{\ominus}$ is independent of temperature and use the integrated form of the van't Hoff isochore:

$$\ln K_p(T_2) = \ln K_p(T_1) - \frac{\Delta_r H^{\ominus}}{R}\left[\frac{1}{T_2} - \frac{1}{T_1}\right]$$

At $T = 198$ K:

$$\ln K_p(198) = \ln K_p(298) - \frac{57,200}{R}\left[\frac{1}{198} - \frac{1}{298}\right]$$

$$K_p(198) = 1.283 \times 10^{-6}$$

At $T = 398$ K:

$$\ln K_p(398) = \ln K_p(298) - \frac{57,200}{8.314}\left[\frac{1}{398} - \frac{1}{298}\right]$$

$$K_p(398) = 49.1$$

The degree of dissociation at 298 and 398 K, at pressures of 1 and 10 bar, are then calculated by inserting these values of K_p into the expression for α. The final results are:

T/K	K_p	p/bar	α
198	1.28×10^{-6}	1	5.7×10^{-4}
		10	1.8×10^{-4}
298	0.149	1	0.189
		10	0.061
398	49.1	1	0.962
		10	0.742

Chapter 8

1. In the structure of solid carbon monoxide, all the molecular axes are aligned. Although the energetically favoured structure has all the dipole moments oriented parallel to each other, at finite temperatures there will be disorder with some molecules aligned anti-parallel. This is due to the greater entropy of a structure in which the orientation of the molecules is not uniform, since there are more microstates. This is further assisted by the fact that the dipole moment of CO is small. Schematically, we can picture the two arrangements as:

:C=O :C=O *vs.* :C=O O=C:

When the solid is cooled to absolute zero, the disordered orientations are frozen in, since the molecules do not have sufficient energy to overcome the barrier to rotation. Hence the entropy fails to go to zero. It is generally true that systems can be kinetically trapped in metastable disordered states. Many everyday materials, such as glass for example, should exist in a different form under ambient conditions based on thermodynamic stability alone. We can go one step further in this problem and use the value of S given in the question, together with equation (8.1), to evaluate the number of microstates, W. This yields $W \approx 2$, which is consistent with the schematic picture of two possible relative orientations of neighbouring molecules.

2. Without considering the order of occupancies of the levels or which particular levels are involved, since neither of these influence the answer, we can see that the possible configurations are (4,0,0,0), (3,1,0,0), (2,2,0,0), (2,1,1,0) and (1,1,1,1). For these the values of W would be 30, 120, 180, 360 and 720, respectively. Hence placing each particle in a separate level would maximize the entropy. The reason that particles do not always adopt this configuration is due to the internal energy, which would tend to cause all particles to exist in the ground state. If the separation of the levels is uniform, then the energies of the above five configurations in units of the separation, relative to the lowest level, would be 0, 1, 2, 3 and 6, respectively. Thus the trend in the increasing entropy is the same as that for the internal energy. Hence the optimal configuration is a compromise, between these two competing trends, that minimizes the free energy at any given temperature.

3. Cancelling 5! against part of 10! we can simplify the number of microstates before evaluating:

$$W = \frac{10!}{5!\,3!\,2!0!} = \frac{10 \times 9 \times 8 \times 7 \times 6}{3 \times 2 \times 2} = 10 \times 9 \times 4 \times 7 = 2520$$

$$S = k_{\mathrm{B}} \ln W = 1.08 \times 10^{-22}\ \mathrm{J\ K^{-1}} \quad \text{(to 3 s.f.)}$$

4. By exciting a single particle from the configuration (5,3,2,0), we can arrive at the following three possibilities: (4,4,2,0), (5,2,3,0) and (5,3,1,1). If we call W_0 the number of microstates for (5,3,2,0), then the number for each of the other states can be evaluated relative to this as $5W_0/4$, W_0 and $2W_0$. Hence the occupancy configuration that would maximize the entropy is (5,3,1,1), since this maximizes the number of microstates and gives an entropy of 1.177×10^{22} J.

The Helmholtz free energy difference between this state and the original one can be calculated using the analogous relationship to equation (7.16) under conditions of fixed volume. Using this we obtain:

$$\Delta A = \Delta U - T\Delta S = 1 \times 10^{-20} - 298 \times \left(1.177 - 1.081\right) \times 10^{-22}\ \mathrm{J}$$

$$= +0.714 \times 10^{-20}\,\mathrm{J} \quad \text{(to 3 s.f.)}$$

Chapter 9

1. (a) There are a total of 11 distinct configurations that are possible. Using the notation of Chapter 1, where the occupancies are listed for each level commencing with the ground state, the configurations are:

1	(7,0,0,0,0,0,1)	$W = 8$
2	(6,1,0,0,0,1,0)	$W = 56$
3	(6,0,1,0,1,0,0)	$W = 56$
4	(6,0,0,2,0,0,0)	$W = 28$
5	(5,2,0,0,1,0,0)	$W = 168$
6	(5,1,1,1,0,0,0)	$W = 336$
7	(5,0,3,0,0,0,0)	$W = 112$
8	(4,3,0,1,0,0,0)	$W = 560$
9	(4,2,2,0,0,0,0)	$W = 420$
10	(3,4,1,0,0,0,0)	$W = 560$
11	(2,6,0,0,0,0,0)	$W = 28$

(b) The number of microstates is listed for each configuration above. The state(s) with the greatest number of microstates will maximize the entropy and thus lead to the lowest free energy. Hence we can see that there are two states which will be equally most probable, both having 560 microstates:

(4,3,0,1,0,0,0) and (3,4,1,0,0,0,0)

2. Assuming the number of particles to be large, we can use the Boltzmann distribution to evaluate the percentage in each state. The first step is to evaluate the exponential factor for each state, noting that because the energy differences are given in kJ mol^{-1} we need to use R (8.314 J K^{-1} mol^{-1}) instead of k_B:

$$n_0 \propto \exp\left(-\frac{0.0}{RT}\right) = 1$$

$$n_1 \propto \exp\left(-\frac{0.72}{8.314\times10^{-3}\times300}\right) = 0.74926$$

$$n_2 \propto \exp\left(-\frac{1.24}{8.314\times10^{-3}\times300}\right) = 0.60826$$

To obtain the percentage, P, in each level, we divide the factor for each level by the sum over the levels, multiplied by 100%:

$$P_0 = \frac{1.0}{1.0+0.74926+0.60826}\times100\% = 42.4\% \quad \text{(to 3 s.f.)}$$

$$P_1 = \frac{0.74926}{2.35752}\times100\% = 31.8\% \quad \text{(to 3 s.f.)}$$

$$P_2 = \frac{0.60826}{2.35752}\times100\% = 25.8\% \quad \text{(to 3 s.f.)}$$

Chapter 10

1. To perform the differentiation we must use the quotient rule to handle the fraction, followed by the chain rule to deal with the individual terms:

$$C_V = \frac{\frac{\partial}{\partial T}\left(\sum_i U_i \exp(-U_i/k_BT)\right)}{\sum_i \exp(-U_i/k_BT)} - \frac{\left(\sum_i U_i \exp(-U_i/k_BT)\right)}{\left(\sum_i \exp(-U_i/k_BT)\right)^2}\frac{\partial}{\partial T}\left(\sum_i \exp(-U_i/k_BT)\right)$$

$$= \frac{\sum_i \frac{U_i^2}{k_B T^2} \exp(-U_i/k_B T)}{\sum_i \exp(-U_i/k_B T)} - \frac{\left(\sum_i U_i \exp(-U_i/k_B T)\right)}{\left(\sum_i \exp(-U_i/k_B T)\right)^2}\left(\sum_i \frac{U_i}{k_B T^2} \exp(-U_i/k_B T)\right)$$

$$= \frac{1}{k_B T^2}\left(\frac{\sum_i U_i^2 \exp(-U_i/k_B T)}{\sum_i \exp(-U_i/k_B T)} - \left(\frac{\sum_i U_i \exp(-U_i/k_B T)}{\sum_i \exp(-U_i/k_B T)}\right)^2\right)$$

The first and second terms in the outer bracket can be identified as and by definition. Hence we arrive at the desired result:

$$C_V = \frac{\langle U^2 \rangle - \langle U \rangle^2}{k_B T^2}$$

2. Here the key task is to manipulate the numerator. Starting from the definition of δU and taking its square:

$$(\delta U)^2 = (U - \langle U \rangle)^2 = U^2 - 2U\langle U \rangle + \langle U \rangle^2$$

Now taking the average of this quantity leads to:

$$\langle (\delta U)^2 \rangle = \langle U^2 \rangle - 2\langle U \rangle \langle U \rangle + \langle U \rangle^2 = \langle U^2 \rangle - \langle U \rangle^2$$

Substituting for the numerator in the result of the previous problem yields the desired result:

$$C_V = \frac{\langle (\delta U)^2 \rangle}{k_B T^2}$$

Chapter 11

1. For a solid the total partition function for four identical atoms would be q^4 where q is the atomic partition function. However, when melting occurs the atoms are no longer distinguishable, and thus the total partition function is now $q^4/4!$ Hence the total partition function will *decrease* on melting by a factor of $4! = 24$.

2. (a) For the separate systems, the partition functions can be written down remembering that only like atoms are indistinguishable:

$$Q_1 = \frac{q_A^4 q_B^2}{4!\,2!} = \frac{q_A^4 q_B^2}{48} \qquad Q_2 = \frac{q_A^3 q_B^3}{3!\,3!} = \frac{q_A^3 q_B^3}{36}$$

(b) When the atoms are combined into one container, the result is that for a system consisting of 7 atoms of A and 5 of B:

$$Q_{1+2} = \frac{q_A^7 q_B^5}{7!5!} = \frac{q_A^7 q_B^5}{604800}$$

(c) This part is more subtle: by linking the containers together the volume of the system has been doubled. Given that the translational partition function per atom is proportional to the volume, the values of q_A and q_B will therefore be double their values for the original system. Hence in terms of the original atomic partition functions we obtain:

$$Q_{1+2} = \frac{(2q_A)^7 (2q_B)^5}{7!5!} = \frac{32 q_A^7 q_B^5}{4725}$$

3. In the following parts, remember to use only SI units. (a) The de Broglie wavelength is given by:

$$\lambda = \left(\frac{h^2}{2\pi m k_B T}\right)^{1/2} = \left(\frac{(6.6262 \times 10^{-34}\ \text{J s})^2}{2\pi\left(\frac{79.90 \times 10^{-3}}{6.022 \times 10^{23}}\ \text{kg}\right)(1.38066 \times 10^{-23}\ \text{J K}^{-1})(500\ \text{K})}\right)^{1/2}$$

$$= 8.735 \times 10^{-12}\ \text{m} \quad \text{(to 3 d.p.)}$$

(b) To calculate the translational partition function we can use the de Broglie wavelength:

$$q^T = \lambda^{-3} V = (8.735 \times 10^{-12}\ \text{m})^{-3} (35 \times 10^{-3}\ \text{m}^3) = 5.25 \times 10^{31} \quad \text{(to 3 s.f.)}$$

4. There are two approaches to solving this problem: either by using the expressions already seen for U and S, or by direct derivation from the definition of the Helmholtz free energy in terms of the total partition function. As the latter is the more complex, it is the one that will be presented here, though of course the answers are identical.

$$A = -k_B T \ln Q = -k_B T \ln\left(\frac{q^N}{N!}\right) = -N k_B T \ln q + k_B T \ln N!$$

At this point it is necessary to introduce Stirling's approximation:

$$A = -N k_B T \ln q + N k_B T \ln N - N k_B T = -N k_B T \ln\left(\frac{q}{N}\right) - N k_B T$$

Now we can introduce the specific expression for the translational partition function to yield the final result:

$$A = -Nk_BT\ln\left[\frac{(2\pi mk_BT)^{3/2}}{Nh^3}V\right] - Nk_BT$$

5. The solution here is to use the Sackur–Tetrode equation (11.39), taking care to convert all quantities to SI units:

$$S_m = \frac{5R}{2} + R\ln\left[\frac{\left(2\pi\left(\frac{65.38\times10^{-3}}{N_A}\right)k_B(800)\right)^{3/2}}{N_Ah^3}\frac{R(800)}{10^9}\right]$$

$$= 20.785 + 8.314\ln\left[\frac{(7.5346\times10^{-45})^{3/2}}{1.752\times10^{-76}}6.6512\times10^{-6}\right]\text{J K}^{-1}\text{ mol}^{-1}$$

$$= 20.785 + 8.314\ln\left[2.4829\times10^{4}\right] = 104.9\text{ J K}^{-1}\text{ mol}^{-1} \quad \text{(to 1 d.p.)}$$

Chapter 12

1. (a) The pressure is calculated by taking the volume derivative of the partition function. However, the rotational partition function does not depend on the volume. Therefore the rotational contribution to the pressure is zero.
(b) To derive an expression for the entropy, begin by writing the definition in terms of the total partition function, and then introduce the rotational partition function:

$$S = k_BT\left(\frac{\partial\ln Q}{\partial T}\right)_V + k_B\ln Q \qquad Q = \left(\frac{T}{\sigma\theta_r}\right)^N$$

$$\therefore S = Nk_BT\left(\frac{\partial\ln T}{\partial T}\right)_V + Nk_B\ln\left(\frac{T}{\sigma\theta_r}\right) = Nk_B + Nk_B\ln\left(\frac{T}{\sigma\theta_r}\right)$$

Note that the entropy is one of the thermodynamic properties where the symmetry number does influence the final result.

2. The first stage is to calculate the moment of inertia in the SI units of kg m^2. Note that the process is simplified by using the fact that $m_1 = m_2$:

$$I = \left(\frac{m_1 m_2}{m_1 + m_2}\right) r^2 = \left(\frac{35.453 \times 10^{-3}}{2 \times 6.022 \times 10^{23}}\right)\left(0.1986 \times 10^{-9}\right)^2 \text{kg m}^2$$
$$= 1.161 \times 10^{-45} \text{kg m}^2$$

This result can then be combined with the remaining fundamental constants to give the rotational temperature:

$$\theta_r = \frac{h^2}{8\pi^2 I k_B} = \frac{\left(6.6262 \times 10^{-34} \text{ J s}\right)^2}{8\pi^2\left(1.161 \times 10^{-45} \text{ kg m}^2\right)\left(1.38066 \times 10^{-23} \text{ J K}^{-1}\right)} = 0.347 \text{ K}$$

Note that the symmetry number, which for Cl_2 would be 2, is not included in the rotational temperature. Using equation (12.16), we can now calculate the rotational partition function:

$$q^R = \frac{T}{\sigma\theta_r} = \frac{298.15}{2 \times 0.347} = 429.61 \qquad \text{(to 2 d.p.)}$$

3. First convert the frequency into Hz (s^{-1}) by multiplying by the speed of light (in cm s^{-1}):

$$\omega = 561.1 \times 2.9979 \times 10^{10} \text{ Hz} = 1.682 \times 10^{13} \text{ Hz}$$

From this we can calculate the vibrational temperature:

$$\theta_v = \frac{h\omega}{k_B} = \frac{6.6262 \times 10^{-34} \times 1.682 \times 10^{13}}{1.38066 \times 10^{-23}} \text{K} = 807.24 \text{ K}$$

Equation (12.40) can now be used to arrive at the internal energy of vibration, using $N = 1$ for a single molecule:

$$U^V = \frac{6.6262 \times 10^{-34} \times 1.682 \times 10^{13}}{2} + \frac{1.38066 \times 10^{-23} \times 807.24}{\left(\exp\left(\frac{807.24}{298}\right) - 1\right)} \text{J}$$
$$= 5.573 \times 10^{-21} + 7.954 \times 10^{-22} \text{ J} = 6.368 \times 10^{-21} \text{ J} \qquad \text{(to 3 d.p.)}$$

The classical limit for the vibrational energy based on equation (12.43), but for a single diatomic molecule rather than a mole, is:

$$U^V_{\text{classical}} = k_B T = 1.38066 \times 10^{-23} \times 298 \text{ J} = 4.114 \times 10^{-21} \text{ J} \qquad \text{(to 3 d.p.)}$$

Comment: At first sight it appears odd that the quantum internal energy of vibration from statistical mechanics is higher than the classical internal energy from the equipartition theorem, given the fact that quantization restricts the uptake of energy until higher temperatures. The reason for this is the zero point energy of vibration, which is the dominant contribution to the quantized result, and is unaccounted for in the classical internal energy.

4. For a single harmonic oscillator $Q = q^{\mathrm{V}}$. Hence we can substitute the expression for the vibrational partition function of a harmonic oscillator into the general equation for the entropy:

$$S = k_{\mathrm{B}}T\left(\frac{\partial \ln q^{\mathrm{V}}}{\partial T}\right)_V + k_{\mathrm{B}}\ln q^{\mathrm{V}} \qquad q^{\mathrm{V}} = \frac{\exp(-\theta_{\mathrm{V}}/2T)}{1-\exp(-\theta_{\mathrm{V}}/T)}$$

$$\therefore S = k_{\mathrm{B}}T\left[\frac{\partial}{\partial T}\left(-\frac{\theta_{\mathrm{V}}}{2T} - \ln\left(1-\exp\left(-\frac{\theta_{\mathrm{V}}}{T}\right)\right)\right)\right] + k_{\mathrm{B}}\left[-\frac{\theta_{\mathrm{V}}}{2T} - \ln\left(1-\exp\left(-\frac{\theta_{\mathrm{V}}}{T}\right)\right)\right]$$

$$S = \frac{k_{\mathrm{B}}\theta_{\mathrm{V}}}{2T} + \frac{1}{\left(1-\exp(-\theta_{\mathrm{V}}/T)\right)}\frac{k_{\mathrm{B}}\theta_{\mathrm{V}}}{T}\exp\left(-\frac{\theta_{\mathrm{V}}}{T}\right) - \frac{k_{\mathrm{B}}\theta_{\mathrm{V}}}{2T} - k_{\mathrm{B}}\ln\left(1-\exp\left(-\frac{\theta_{\mathrm{V}}}{T}\right)\right)$$

$$S = \frac{k_{\mathrm{B}}\theta_{\mathrm{V}}\exp(-\theta_{\mathrm{V}}/T)}{T\left(1-\exp(-\theta_{\mathrm{V}}/T)\right)} - k_{\mathrm{B}}\ln\left(1-\exp\left(-\frac{\theta_{\mathrm{V}}}{T}\right)\right)$$

5. The electronic partition function is calculated by a direct sum over the levels, remembering to allow for the degeneracies. Also, the energy separation must be convert to J:

$$q^{\mathrm{E}} = 2\exp(-0) + 2\exp\left(-\frac{139.7\times 2.9979\times 10^{10}\times 6.6262\times 10^{-34}}{1.38066\times 10^{-23}\times 298}\right)$$
$$= 2 + 1.0188 = 3.019 \quad \text{(to 3 d.p.)}$$

6. (a) We can write down the electronic partition function for the two-level system:

$$q^{\mathrm{E}} = 2\exp(0) + 2\exp\left(-\frac{6.6262\times 10^{-34}\times 121.1\times 2.9979\times 10^{10}}{1.38066\times 10^{-23}\times T}\right)$$
$$= 2 + 2\exp\left(-\frac{174.2365}{T}\right)$$

We then need the general expression for the heat capacity, which can be obtained by differentiating that for the internal energy, in terms of the partition function, with respect to temperature at constant volume:

$$C_V = \frac{\partial}{\partial T}\left[k_B T^2 \left(\frac{\partial \ln q^E}{\partial T}\right)_V\right] = 2k_B T\left(\frac{\partial \ln q^E}{\partial T}\right)_V + k_B T^2 \left(\frac{\partial^2 \ln q^E}{\partial T^2}\right)_V$$

$$C_V = \frac{2k_B T}{q^E}\left(\frac{\partial q^E}{\partial T}\right)_V + k_B T^2 \left(\frac{1}{q^E}\left(\frac{\partial^2 q^E}{\partial T^2}\right)_V - \frac{1}{\left(q^E\right)^2}\left(\frac{\partial q^E}{\partial T}\right)_V^2\right)$$

Working out the derivatives of the electronic partition function with respect to temperature, in terms of the electronic temperature ($\theta_e = h\Delta u / k_B$) for generality, before substituting for 174.24 K later:

$$\left(\frac{\partial q^E}{\partial T}\right)_V = \frac{2\theta_e}{T^2}\exp\left(-\frac{\theta_e}{T}\right)$$

$$\left(\frac{\partial^2 q^E}{\partial T^2}\right)_V = \frac{2\theta_e}{T^2}\left(\frac{\theta_e}{T^2} - \frac{2}{T}\right)\exp\left(-\frac{\theta_e}{T}\right)$$

Substituting these expressions into that for the heat capacity at constant volume, and simplifying, yields:

$$C_V = \frac{4k_B\theta_e^2 \exp\left(-\theta_e / T\right)}{\left(q^E\right)^2 T^2}$$

(b) When sketching the form of the heat capacity curve it is necessary to know the limits at the extremes of temperature. As the temperature goes to absolute zero, the exponential term will dominate leading to the heat capacity going to zero. At high temperatures the term $1/T^2$ will also make the heat capacity tend to zero. Given that all the terms in the equation must be positive, we know that the heat capacity curve will not cross the axis. Hence we can deduce that the curve must rise from zero, go through a maximum and then decay to zero again. Hence the heat capacity will have the following appearance:

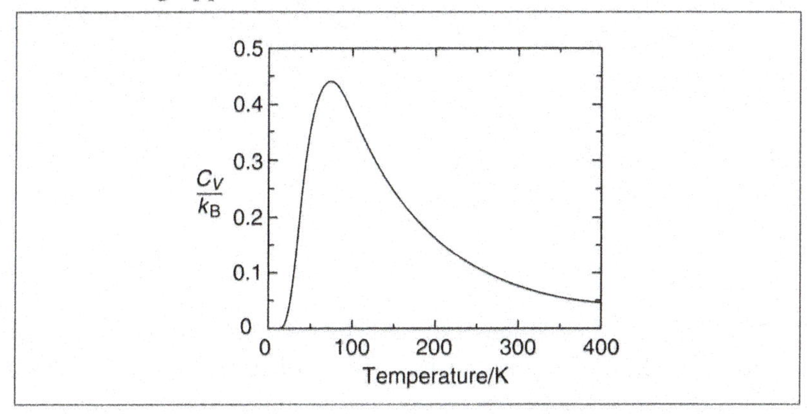

Figure A.1 The electronic heat capacity C_V relative to k_B of a molecule of NO as a function of temperature

To prove that there is a maximum, we would have to demonstrate that there is a point where the first derivative of C_V with respect to temperature is zero and that the curvature is positive at this point:

$$\left(\frac{\partial C_V}{\partial T}\right) = \frac{4k_B\theta_e^2 \exp(-\theta_e/T)}{(q^E)^2 T^3}\left(\frac{\theta_e}{T} - 2 - \frac{4\theta_e}{T}\frac{\exp(-\theta_e/T)}{q^E}\right)$$

When the final term in brackets goes to zero, the heat capacity reaches a stationary point. Unfortunately, rearranging the above equation to make T the subject of the formula is non-trivial, but the solution can be found to be when $T = 0.42\theta_e$. Differentiating again and substituting this value for T demonstrates that the curvature is negative, as required for a maximum.

(c) The reason for the shape is as follows. At low temperatures, quantization means that the heat capacity is initially zero since the higher level is not accessible. As the temperature increases, so electrons become excited to the second level until saturation of the upper level is reached (where the population of the levels is close to equal). At this point the heat capacity falls to zero because there is no means for the additional energy to be absorbed.

Chapter 13

1. The reaction for the formation of HCl is:

$$\frac{1}{2}H_{2(g)} + \frac{1}{2}Cl_{2(g)} \rightleftharpoons HCl_{(g)}$$

Hence the expression for the standard Gibbs free energy of reaction is:

$$\Delta_r G^\ominus = \Delta_r U^\ominus(0) - RT\ln\left[\frac{\left(q^\ominus_{HCl}/N_A\right)}{\left(q^\ominus_{H_2}/N_A\right)^{1/2}\left(q^\ominus_{Cl_2}/N_A\right)^{1/2}}\right]$$

For each reactant and product molecule we know that the molecular partition function can be written as:

$$q = q^T q^R q^V$$

All of the molecules have closed-shell singlet ground states and therefore the electronic partition function is always 1. Hence we have to calculate the individual partition functions for each of the forms of motion for each molecule following the method used in previous problems for each type. The resulting values are:

Molecule	q^T	I (kg m^2)	q^R	q^V
H_2	6.77502×10^{28}	4.60412×10^{-48}	1.7041 ($\sigma = 2$)	2.4493×10^{-5}
Cl_2	5.21139×10^{30}	1.16109×10^{-45}	429.75 ($\sigma = 2$)	0.27670
HCl	1.41331×10^{31}	2.64419×10^{-47}	19.574 ($\sigma = 1$)	7.3662×10^{-4}

Next we need to calculate the internal energy of reaction, using the dissociation energies given. Note that we must also correct for the zero point energies of all molecules by adding these to the dissociation energy:

$$\Delta_r U(0) = -\left(D_0(\mathrm{HCl}) + U_{ZPE}(\mathrm{HCl})\right) + \frac{1}{2}\left(D_0(\mathrm{H_2}) + U_{ZPE}(\mathrm{H_2}) + D_0(\mathrm{Cl_2}) + U_{ZPE}(\mathrm{Cl_2})\right)$$

$$= -92.095 \text{ kJ mol}^{-1} + \frac{1}{2} N_A hc\left(-2989.6 + \frac{1}{2}(4400.4 + 561.1)\right)$$

$$= -92.095 - 3.044 \text{ kJ mol}^{-1} = -95.14 \text{ kJ mol}^{-1} \qquad \text{(to 2 d.p.)}$$

Finally, we can collect the two contributions to the free energy together:

$$\Delta_r G^0 = -95.14 - 8.314 \times 10^{-3} \times 298.15 \ln[4.86796] \text{ kJ mol}^{-1}$$

$$= -95.14 - 3.92 \text{ kJ mol}^{-1} = -99.1 \text{ kJ mol}^{-1}$$

2. The equilibrium constant can be written in terms of the molecular partition functions for the reactants and products as:

$$K_p = \frac{(q_{\mathrm{Li}}/N_A)^2}{q_{\mathrm{Li_2}}/N_A} \exp\left(-\frac{\Delta U(0)}{RT}\right)$$

Factorizing each partition function into the relevant contributions:

$$K_p = \frac{\left(q_{\mathrm{Li}}^{T} q_{\mathrm{Li}}^{E}\right)^2}{N_A\left(q_{\mathrm{Li_2}}^{T} q_{\mathrm{Li_2}}^{R} q_{\mathrm{Li_2}}^{V}\right)} \exp\left(-\frac{\Delta U(0)}{RT}\right)$$

The first step is to evaluate all the partition function contributions under the conditions specified. The calculated values:

Molecule	q^T	q^R	q^V	q^{el}
Li_2	1.22394×10^{30}	152.384 ($\sigma = 2$)	0.52057	1
Li	4.32727×10^{29}	–	–	2

The internal energy of reaction is given by the dissociation energy plus the zero point energy of the Li_2 molecule:

$$\Delta_r U(0) = D_0 + \frac{1}{2} N_A hc\omega$$

$$= 99.38 + \frac{6.022 \times 10^{23} \times 6.6262 \times 10^{-34} \times 2.9979 \times 10^{10} \times 353.59}{2 \times 1000} \text{ kJ mol}^{-1}$$

$$= 101.495 \text{ kJ mol}^{-1}$$

Substituting all values into the expression for K_p:

$$K_p = \frac{7.7145 \times 10^{27}}{6.022 \times 10^{23}} \exp\left(-\frac{101.495}{8.314 \times 10^{-3} \times 298.15} \right)$$

$$= 12810.57 \exp(-40.94) = 2.116 \times 10^{-14}$$

Comment: Lithium molecules show negligible dissociation under standard conditions.

Subject Index

Lightning Source UK Ltd.
Milton Keynes UK
UKOW05f0007050817
306697UK00002B/83/P